METHODS
IN
YEAST GENETICS

A Cold Spring Harbor Laboratory Course Manual

2000 Edition

METHODS
IN
YEAST GENETICS

A Cold Spring Harbor Laboratory Course Manual

2000 Edition

Dan Burke

Dean Dawson

Tim Stearns

COLD SPRING HARBOR LABORATORY PRESS

METHODS IN YEAST GENETICS

A Cold Spring Harbor Laboratory Course Manual
2000 Edition

Project Coordinator:	Inez Sialiano
Production Editors:	Pat Barker, Mala Mazzullo
Interior Designer:	Denise Weiss
Cover Designer:	Koulian Williams Design Group

Front Cover: A culture of *Saccharomyces cerevisiae* incubated in medium containing the viability indicator FUN-1™ and the counterstain Calcofluor™ White M2R, both of which are provided in our LIVE/DEAD™ Yeast Viability Kit (L-7009). Metabolically active yeast process the FUN-1 dye, forming numerous red fluorescent cylindrical structures within their vacuoles. Calcofluor stains the chitin in yeast cell walls fluorescent blue, regardless of the cell's metabolic state.

Photo by Bruce Roth and Paul Millard, ©Molecular Probes, Inc. Calcofluor is a trademark of American Cyanamid. All other trademarks and registered trademarks are owned by Molecular Probes, Inc.

Library of Congress Cataloging-in-Publication Data

Burke, Dan, 1954-
      Methods in yeast genetics : a Cold Spring Harbor Laboratory course manual / by
Dan Burke, Dean Dawson, Tim Stearns.— 2000 ed.
      p. cm.
      Rev. ed. of: Methods in yeast genetics / Alison Adams ... [et al.].
      1997 ed. 1998.
      Includes bibliographical references.
      ISBN 0-87969-588-9
      1. Yeast fungi—Genetics—Laboratory manuals. I. Dawson, Dean. II.
Stearns, Tim. III. Cold Spring Harbor Laboratory. IV. Methods in yeast genetics. V. Title.
QK617.5 .K35 2000
579.5'62135—dc21
                          00-031721

10 9 8 7 6 5 4 3 2 1

All Cold Spring Harbor Laboratory Press publications may be ordered directly from Cold Spring Harbor Laboratory Press, 10 Skyline Drive, Plainview, New York 11803-2500. Phone: 1-800-843-4388 in Continental U.S. and Canada. All other locations: (516) 349-1930. Fax: (516) 349-1946. E-mail: cshpress@cshl.org. For a complete catalog of all Cold Spring Harbor Laboratory Press publications, visit our World Wide Web site: http://www.cshl.org

Contents

Appendices

Preface

This laboratory course manual incorporates significant portions of the manuals used in previous Cold Spring Harbor Yeast Genetics Courses. Although most of the experiments have now been revised and several new techniques have been added, the basic structure of this course is the same as in those that were taught in the Cold Spring Harbor Yeast Genetics Course for the past 30 years. We are indebted to our predecessors, Fred Sherman, Gerry Fink, Jim Hicks, Mark Rose, Fred Winston, Phil Hieter, Susan Michaelis, Aaron Mitchell, Alison Adams, Chris Kaiser, and Dan Gottschling, for their teaching and for making this course an important part of the yeast community. We also thank Mike Cherry for his invaluable assistance with the genetic and physical maps.

Dan Burke
Dean Dawson
Tim Stearns

Introduction

Genetic investigations of yeast were essentially initiated by Winge and his co-workers in the mid-1930s. Approximately ten years later, Lindegren and his colleagues also began extensive studies. These two groups are responsible for uncovering the general principles and much of the basic methodology of yeast genetics. Today, yeast is widely recognized as an ideal eukaryotic microorganism for biochemical and genetic studies. Although yeasts have greater genetic complexity than bacteria, they still share many of the technical advantages that permitted rapid progress in the molecular genetics of prokaryotes and their viruses. Some of the properties that make yeast particularly suitable for genetic studies include the existence of both stable haploid and diploid cells, rapid growth, clonability, the ease of replica plating, mutant isolation, and ability to isolate each haploid product of meiosis by microdissection of a tetrad ascus. Yeast has been successfully employed for the study of all areas of genetics, such as mutagenesis, recombination, chromosome segregation, gene action and regulation, as well as aspects distinct to eukaryotic systems, such as mitochondrial genetics.

DNA transformation has made yeast particularly accessible to gene cloning and genetic engineering techniques. Structural genes corresponding to virtually any genetic trait can be identified by complementation from plasmid libraries. DNA is introduced into yeast cells either as replicating molecules or by integration into the genome. In contrast to most other organisms, integrative recombination of transforming DNA in yeast proceeds primarily via homologous recombination. This permits efficient targeted integration of DNA sequences into the genome. Homologous recombination coupled with high levels of gene conversion has led to the development of techniques for the direct replacement of normal chromosomal loci with genetically engineered DNA sequences. This ease of performing direct gene replacement is unique among eukaryotic organisms and has been extensively exploited in every aspect of yeast genetics, cell biology, physiology, and biochemistry. Many of these modern genetic techniques are reviewed in Volume 194 of *Methods in Enzymology* (Guthrie and Fink 1991) and by Rose (1995).

Recent advances in modern yeast genetics have come from determining the complete DNA sequence of the *Saccharomyces cerevisiae* genome. The assembly of this sequence into a public sequence database has made yeast one of the premier organisms for detailed analysis of eukaryotic cellular function and genome organization. Several Internet web sites provide convenient and organized access to the sequence database.

Two of the more popular sites are the Saccharomyces Genome Database (SDB) and the Munich Information Center for Protein Sequences (MIPS). Each of these sources augments the sequence database by providing powerful search functions for analyzing the genome, as well as helpful links to relational databases, such as literature cross-references or protein structure of yeast gene products. In addition, the Yeast Protein Database (YPD) has each yeast protein and predicted open reading frame compiled with information about its structure and the phenotypes associated with mutations in its gene.

One of the most useful collections of information on *Saccharomyces cerevisiae* biology is found in two companion series of reviews entitled *The Molecular Biology of the Yeast* Saccharomyces (Strathern et al. 1981 and 1982) and *The Molecular and Cellular Biology of the Yeast* Saccharomyces (Broach et al. 1991; Jones et al. 1992; Pringle et al. 1997). While a few chapters in the later series update progress made since the first series was published, each volume has detailed reviews that synthesize vast literatures of yeast biology that are not available anywhere else. All five of these books are essential for the library of a yeast biologist. *The Early Days of Yeast Genetics* is also a book to consider reading, for it provides a historical perspective on why yeast became a model organism and traces the development of the yeast community's influence on modern genetics and eukaryotic molecular biology.

This course will focus exclusively on the baker's yeast, *Saccharomyces cerevisiae*. After completing the course, you should be able to carry out all of the techniques commonly employed by yeast geneticists and be able to follow the literature with greater ease. Except for the dissection of asci, most of the methods do not differ significantly from the methods employed with other microorganisms, and the skills should be rapidly acquired with little practice. Experiments will be conducted in pairs, and whenever possible, an investigator more familiar with microbiological techniques will be assigned to a less-experienced partner. Since most of the experiments will be initiated at the outset of the course, it is advisable to read the entire manual thoroughly. Please note that some experiments span many days due to extended periods of incubation.

We wish to emphasize that some of the procedures in this manual have been condensed in order to save time and are not necessarily standard for research purposes. For example, mutants are usually purified by subcloning the initial isolates. Also, some of the techniques may not be directly applicable to your research problems, but they have been included to illustrate general principles and methods.

HIGHLY RECOMMENDED BOOKS

Broach J.R., Jones E.W., and Pringle J.R. 1991. *The molecular and cellular biology of the yeast* Saccharomyces. I. *Genome dynamics, protein synthesis, and energetics.* Cold Spring Harbor Laboratory Press, Cold Spring Harbor, New York.

Guthrie C. and Fink G.R., eds. 1991. Guide to yeast genetics and molecular biology. *Methods Enzymol.*, vol. 194.

Jones, E.W., Pringle J.R., and Broach J.R. 1992. *The molecular and cellular biology of the yeast* Saccharomyces. II. *Gene expression.* Cold Spring Harbor Laboratory Press, Cold Spring Harbor, New York.

Pringle J.R., Broach J.R., and Jones E.W. 1997. *The molecular and cellular biology of the yeast* Saccharomyces. III. *Cell cycle and cell biology.* Cold Spring Harbor Laboratory Press, Cold Spring Harbor, New York.

Strathern, J.N., E.W. Jones, and J.R. Broach, eds. 1981. *The molecular biology of the yeast* Saccharomyces: *Life cycle and inheritance.* Cold Spring Harbor Laboratory, Cold Spring Harbor, New York.

———. 1982. *The molecular biology of the yeast* Saccharomyces: *Metabolism and gene expression.* Cold Spring Harbor Laboratory, Cold Spring Harbor, New York.

OTHER BOOKS ABOUT YEAST

Fincham J.R.S., Day P.R., and Radford A. 1979. *Fungal genetics.* University of California Press, Berkeley and Los Angeles.

Rose M.D. 1995. Modern and post-modern genetics in *Saccharomyces cerevisiae.* In *The yeasts,* 2nd edition (ed. A.E. Wheals et al.), vol. 6, pp. 69–120. Academic Press, New York.

Hall M.N. and Linder P. 1993. *The early days of yeast genetics.* Cold Spring Harbor Laboratory Press, Cold Spring Harbor, New York.

RECOMMENDED YEAST WEB SITES

Munich Information Center for Protein Sequences
http://speedy.mips.biochem.mpg.de/mips/yeast/

Saccharomyces Genome Database
http://genome-www.stanford.edu/Saccharomyces/

Yeast Protein Database
http://www.proteome.com/YPDhome.html

Genetic Nomenclature

CHROMOSOMAL GENES

Early recommendations for the nomenclature and conventions used in yeast genetics have been summarized by Sherman and Lawrence (1974) and Sherman (1981). Whenever possible, gene symbols are consistent with the proposals of Demerec et al. (1966) and are designated by three italicized letters (e.g., *arg*). Contrary to the proposals of Demerec et al. (1966), the genetic locus is identified by a number (not a letter) following the gene symbol (e.g., *arg2*). Dominant alleles are denoted by using uppercase italics for all three letters of the gene symbol (e.g., *ARG2*). Lowercase letters symbolize the recessive allele (e.g., the auxotroph *arg2*). Wild-type genes are designated with a superscript plus sign (*sup6$^+$* or *ARG2$^+$*). Alleles are designated with a number separated from the locus number by a hyphen (e.g., *arg2-14*). Locus numbers are consistent with the original assignments; however, allele numbers may be specific to a particular laboratory.

Phenotypic designations are sometimes denoted by cognate symbols in Roman type followed by a superscript plus or minus sign. For example, the independence from and requirement for arginine can be denoted by Arg$^+$ and Arg$^-$, respectively.

The following examples illustrate the conventions used in the genetic nomenclature for *S. cerevisiae*:

ARG2	A locus or dominant allele
arg2	A locus or recessive allele that produces a requirement for arginine as the phenotype
ARG2$^+$	The wild-type allele of this gene
arg2-9	A specific allele or mutation at the *ARG2* locus
Arg$^+$	A strain that does not require arginine
Arg$^-$	A strain that requires arginine
Arg2p	Designation for the protein product of the *ARG2* gene

There are a number of exceptions to these general rules. Gene clusters, complementation groups within a gene, or domains within a gene that have different properties can be designated by capital letters following the locus number (e.g., *his4A, his4B*).

The extensive use in yeast of recombinant DNA techniques has introduced a nomenclature that pertains to gene insertions, gene fusions, and plasmids:

ARG2::LEU2	An insertion of the *LEU2* gene at the *ARG2* locus where the insertion does not disrupt *ARG2* function
arg2::LEU2	An insertion of the *LEU2* gene at the *ARG2* locus where the insertion disrupts *ARG2* function
arg2-101::LEU2	An insertion of the *LEU2* gene at the *ARG2* locus where the insertion disrupts *ARG2* function and the disruption allele is specified
cyc1–arg2	A gene fusion between the *CYC1* gene and ARG2 where neither gene is functional
P_{cyc1}–*ARG2*	A gene fusion between the *CYC1* gene promoter and *ARG2* where the *ARG2* gene is functional
[YCp–*ARG2*]	A centromere plasmid carrying a functional *ARG2* locus
[pCK101]	Designation for a specific plasmid whose structure is given elsewhere

Although superscripts should be avoided, it is sometimes expedient to distinguish genes conferring resistance or sensitivity by a superscript R or S, respectively. For example, the genes controlling resistance to canavanine sulfate (*can1*) and copper sulfate (*CUP1*) and their sensitive alleles can be denoted, respectively, as *can^{R}1*, *CUP^{R}1*, *CAN^{S}1*, and *cup^{S}1*.

Wild-type and mutant alleles of the mating-type and related loci do not follow the standard rules. The two wild-type alleles of the mating-type locus are designated *MAT*a and *MAT*α. The two complementation groups of the *MAT*α locus are denoted *MAT*α1 and *MAT*α2. Mutations of the *MAT* genes are denoted, e.g., *mat*a-1, *mat*α1-1. The wild-type homothallic alleles at the HMR and HML loci are denoted *HMR*a, *HMR*α, *HML*a, and *HML*α. Mutations at these loci are denoted, e.g., *hmr*a-1, *hml*α-1. The mating phenotypes of *MAT*a and *MAT*α cells are denoted simply **a** and α, respectively.

Dominant and recessive suppressors should be denoted, respectively, by three uppercase or three lowercase letters followed by a locus designation (e.g., *SUP4*, *SUF1*, *sup35*, *suf11*). In some instances, UAA suppressors and UAG suppressors are further designated o and a, respectively, following the locus. For example, *SUP4*-o refers to suppressors of the *SUP4* locus that insert tyrosine residues at UAA sites; *SUP4*-a refers to suppressors of the same *SUP4* locus that insert tyrosine residues at UAG sites. The corresponding wild-type locus coding for the normal tyrosine tRNA and lacking suppressor activity can be referred to as *sup4$^+$*. Thus, the nomenclature describing suppressor and wild-type alleles in yeast is unrelated to the bacterial nomenclature. For example, an ochre *E. coli* suppressor that inserts tyrosine residues at both UAA and UAG sites is denoted as *su$_4^+$*, and the wild-type locus coding for the normal tyrosine tRNA and lacking suppressor activity can be referred to as *Su$_4$*, *su$_4^-$*, or *supC*.

For most structural genes that code for proteins, the functional wild-type allele is usually dominant to the mutant form of a gene. In yeast, the convention for dominant genes utilizes italic symbols such as *HIS4* and *LEU2*. Because the sites of recessive mutations are usually used for genetic mapping, published chromosome maps usually contain the mutant form of the gene. For example, chromosome III contains *his4* and *leu2*, whereas chromosome IX contains *SUP22* and *FLD1*. Because capital letters are used to represent dominant wild-type genes that control the same character (e.g., *SUC1, SUC2*), and because the dominant forms are used in genetic mapping, such chromosomal loci are denoted in capital letters on genetic maps. In addition, capital letters are used to designate certain DNA segments whose locations have been determined by a combination of recombinant DNA techniques and classical mapping procedures (e.g., *RDN1*, the segment encoding ribosomal RNA).

NON-MENDELIAN DETERMINANTS

Where necessary, non-Mendelian genotypes can be distinguished from chromosomal genotypes by enclosure in brackets. Whenever applicable, it is advisable to employ the above rules for designating non-Mendelian genes and to avoid the use of Greek letters. However, when referring to an entire non-Mendelian element, it is best to either retain the original symbols [ρ⁺], [ρ⁻], [ψ⁺], and [ψ⁻] or use their transliteration, [*rho⁺*], [*rho⁻*], [*PSI⁺*], and [*psi⁻*], respectively. Detailed designations for mitochondrial mutants have been presented by Dujon (1981) and Grivell (1984, 1990) and for killer strains by Wickner (1981). Unlike the other non-Mendelian determinants, [*PSI⁺*] and [*URE3*] are not based on different states of a nucleic acid; rather, the [*PSI⁺*] and [*URE3*] traits result from heritable conformational states of proteins. The unusual behavior of these traits can be explained by the prion hypothesis, which has been used to explain infectious neurodegenerative diseases such as scrapie in mammals (Lindquist 1997). [*PSI⁺*] corresponds to a heritable conformational state of the translation termination factor Sup35p, and [*URE3*] corresponds to a heritable inactive state of *URE2*, a gene whose product is involved in nitrogen regulation. The known non-Mendelian determinants in yeast are listed in Table 1.

Table 1. *Non-Mendelian determinants of yeast*

Wild type	Mutant variant	Element	Mutant trait
[ρ⁺]	[ρ⁻]	Mitochondrial DNA	Respiration deficiency
[*KIL*-k₁]	[*KIL*-o]	RNA plasmid	Sensitive to killer toxin
[*cir⁺*]	[*cir⁰*]	2μ plasmid	None
[*psi⁻*]	[*PSI⁺*]	Prion form of Sup35p	Enhanced suppression of non-sense codons
[*ure3⁻*]	[*URE3*]	Prion form of Ure2p	Unregulated ureidosuccinate uptake

GENETIC BACKGROUNDS

The genetic background from which a *S. cerevisiae* strain is derived is an often hidden aspect of the genotype that should be taken into account when designing experiments. Most strains used in modern genetic studies come from one of a small set of genetic backgrounds, including S288C, X2180, A364A, W303, Σ1278b, AB972, SK1, and FL100. The genealogies of some of these backgrounds have recently been reconstructed from records of crosses that were carried out in the 1940s between wild yeasts and brewing strains (Mortimer and Johnston 1986). This analysis shows that although most backgrounds share a common ancestry, a significant degree of genetic heterogeneity has been introduced by outcrossing. In practice, crosses between distantly related strains often give inviable combinations of alleles leading to many inviable spores, whereas crosses between strains from the same background usually give >95% viable spores. The S288C and A364A genetic backgrounds have similar genealogies and in crosses give a high frequency of spore viability, but an analysis of genomic sequences from these strains reveals an average of 3.4 nucleotide sequence differences per kilobase of genomic DNA. Thus, even apparently closely related strains can differ at a very large number of sites.

Allelic differences between strain backgrounds can seriously influence the outcome of many different kinds of experiments, and it is best to avoid genetic heterogeneity as much as possible by using a single genetic background. At the beginning of a new mutant hunt, it is worth considering which strain background to use—usually the background used by most investigators in the same field is the best choice. S288C is probably the most commonly used background; however, other backgrounds offer distinct advantages for particular types of experiments. For example, Σ1278b will form pseudohyphae whereas S288C will not, and SK1 sporulates much more rapidly than S288C. It is often necessary to move a desired mutation from one background into another. Ideally, this can be done using recombinant plasmids and the methods for gene replacement described in Experiment VII. Mutations that have not been cloned can be moved by backcrossing to the desired strain background (usually successive backcrosses are carried out until a clear 2:2 pattern of segregation for the desired trait has been achieved).

REFERENCES

Demerec M., Adelberg E.A., Clark A.J., and Hartman P.E. 1966. A proposal for a uniform nomenclature in bacterial genetics. *Genetics* **54:** 61–76.

Dujon B. 1981. Mitochondrial genetics and functions. In *The molecular biology of the yeast* Saccharomyces: *Life cycle and inheritance* (ed. J.N. Strathern et al.), pp. 505–635. Cold Spring Harbor Laboratory, Cold Spring Harbor, New York.

Grivell L.A. 1984. Restriction and genetic maps of yeast mitochondrial DNA. In *Genetic maps*, 3rd edition (ed. S.J. O'Brien), pp. 234–247. Cold Spring Harbor Laboratory, Cold Spring Harbor, New York.

———. 1990. Mitochondrial DNA in the yeast *Saccharomyces cerevisiae*. In *Genetic maps*, 5th edition (ed. S.J. O'Brien), pp. 3.50–3.57. Cold Spring Harbor Laboratory Press, Cold Spring Harbor, New York.

Lindquist S. 1997. Mad cows meet psi-chotic yeast: The expansion of the prion hypothesis. *Cell* **89**: 495–498.

Mortimer R.K. and Johnston J.R. 1986. Genealogy of principal strains of the yeast genetic stock center. *Genetics* **113**: 35–43.

Sherman F. 1981. Genetic nomenclature. In *The molecular biology of the yeast* Saccharomyces: *Life cycle and inheritance* (ed. J.N. Strathern et al.), pp. 639–640. Cold Spring Harbor Laboratory, Cold Spring Harbor, New York.

Sherman F. and Lawrence C.W. 1974. *Saccharomyces*. In *Handbook of genetics: Bacteria, bacteriophages, and fungi* (ed. R.C. King), vol. 1, pp. 359–393. Plenum Press, New York.

Wickner R.B. 1981. Killer systems in *Saccharomyces cerevisiae*. In *The molecular biology of the yeast* Saccharomyces: *Life cycle and inheritance* (ed. J.N. Strathern et al.), pp. 415–444. Cold Spring Harbor Laboratory, Cold Spring Harbor, New York.

Looking at Yeast Cells

Yeast cells are approximately 5 μm in diameter, and many of their important features can be seen in the light microscope. It is good laboratory practice to routinely examine cultures under phase microscopy for indications of the physiological state of the cells, and for evidence of contamination. Much of modern yeast cell biological work involves more sophisticated examination of yeast cells stained with protein-specific antibodies, or with fluorescent dyes that specifically associate with certain organelles. This experiment will provide examples of the standard types of light microscopy that are used in the examination of yeast cells.

EXAMINATION OF GROWING CULTURES

Growth Properties

Saccharomyces cerevisiae cells grow by budding. A cell that gives rise to a bud is called a mother cell, and the bud is sometimes referred to as the daughter cell. A new bud emerges from a mother cell close to the beginning of the cell cycle and continues to grow throughout the cell cycle until it separates from the mother cell at the end of the cell cycle. Because all of the growth of a yeast cell is concentrated in the bud, and because this growth is essentially continuous throughout the cell cycle, the size of the bud gives an approximate indication of the position of a given cell in the cell cycle. An exponentially growing culture of yeast cells has approximately one-third unbudded cells, one-third cells with a small bud, and one-third cells with a large bud. When cells in a growing culture use up the available nutrients, they stop growing by arresting in the cell cycle as unbudded cells. Thus, a simple way of determining the growth state of a culture is to determine the frequency of budded cells in the microscope. Note that for some strains, the mother and daughter cells remain stuck together even though they have completed cytokinesis. In these cases, it is necessary to vortex or sonicate the culture to separate cells prior to microscopy. Many kinds of mutants also arrest in the cell cycle in a way that is diagnostic of their phenotype. For example, cells in which there is a defect in the mitotic spindle arrest as large budded cells, a point in the cycle that would normally correspond to mitosis. It is important to note that the arrest point, or terminal phenotype, of mutant cells can be morphologically distinct from any cell type seen in a normal culture. In the mitotic mutant above, the mother and daughter cells continue to grow at the arrest point until both are much larger than normal yeast cells.

Haploids vs. Diploids

Haploid and diploid yeast cells are morphologically similar but differ in several important ways. First, diploid cells are larger than haploid cells. Cytoplasmic volume increases with ploidy, and the diameter of a diploid cell is roughly 1.3 times that of a haploid cell. This difference can be readily seen when haploids and diploids are compared side by side. Because they are larger, diploid cells (or even tetraploids in some cases) are often used for fluorescence microscopy where the larger size helps in being able to resolve small cellular structures. Second, diploid cells tend to have a more elongated, or ovoid, shape than haploid cells, which are often almost round. Third, diploids and haploids have a different budding pattern. Yeast cells generally bud about 20 times before becoming senescent. Successive buds emerge from the surface of the mother cell in stereotyped patterns. Haploid cells bud in an axial pattern wherein each bud emerges adjacent to the site of the previous bud. Diploid cells bud in a polar pattern wherein successive buds can emerge from either end of the elongated mother cell. The history of a cell's budding pattern can be visualized by staining cells with Calcofluor, a fluorescent compound that binds to the rings of chitin that remain at old bud sites. These chitin rings are called bud scars, and we will use Calcofluor staining of haploid and diploid cells to visualize the axial and polar bud scar patterns.

Mating Cells

Yeast cells come in three mating types: *MAT*a and *MAT*α; these two are able to mate with each other to yield a *MAT*a/α. *MAT*a/α cells cannot mate with cells of either mating type. Generally, *MAT*a and *MAT*α strains will be haploids and MATa/α strains will be diploid, although this is not always the case, and one should be careful to consider mating type independent of ploidy. The mating process between two cells begins with an exchange of pheromones that causes each of the cells to arrest in the cell cycle as unbudded cells and to induce the expression of proteins required for mating. The pheromone also causes the cells to make a projection of new cell surface specialized for cell fusion. This projection is usually oriented toward the mating partner. Cells with a mating projection are called ''shmoos'' because of their resemblance to an Al Capp cartoon character from the 1940s. The shmooing cells join at the tips of their projections, their cytoplasms fuse, and then their nuclei fuse to form a diploid *MAT*a/α nucleus. The process of nuclear fusion is termed karyogamy. The newly formed diploid is termed a zygote and has a characteristic appearance that is particularly easy to identify when the first bud emerges. It is possible to isolate zygotes by micromanipulation, allowing for the isolation of diploid cells even in situations where there is no genetic selection for diploid formation. We will look at a population of mating cells to identify shmoos and zygotes.

Mitochondria

Yeast mitochondria contain a genome that encodes several proteins involved in oxidative phosphorylation, one ribosomal protein, and the rRNAs and tRNAs required for the mitochondrial translation apparatus. The vast majority of mitochondrial proteins are encoded by nuclear genes and imported into the mitochondria from the cytoplasm. Thus, mutations in either the nuclear genome or the mitochondrial genome can affect mitochondrial function. Yeast cells with mutations in the mitochondrial DNA cannot carry out oxidative phosphorylation, so must get all of their energy from fermentation. Such mutants grow more slowly than wild-type cells and are unable to grow on non-fermentable carbon sources such as lactate, glycerol, or ethanol. The French scientists who first characterized mitochondrial mutants called this the "petite" phenotype. Petite strains form small milky-white colonies. This lack of colony pigmentation is most evident in an *ade2* background; formation of the red pigment that typifies *ade2* mutants requires oxidative phosphorylation. Diploid petite strains are also unable to sporulate, and it is wise to check a nonsporulating strain for the ability to grow on a nonfermentable carbon source before other potential causes of sporulation failure are examined. Petite mutants appear with high frequency in many common lab yeast strains—for some strains, as many as 10% of the cells in a culture are petite. Although the petite phenotype can be due to mutations in either the mitochondrial or nuclear genomes, the great majority of petites are due to mitochondrial DNA mutations. The mitochondrial genome is given the designation "rho"; wild-type strains are rho$^+$, strains with deleted versions of the mitochondrial genome (the most common type of mutation) are rho$^-$, and strains lacking the mitochondrial genome entirely are rho^O. A common misconception is that rho^O strains lack mitochondria altogether. Several essential reactions take place within the mitochondrial membrane, and even in rho^O strains, a diminished mitochondrial structure can be seen in the electron microscope. In this experiment, we will use fluorescence microscopy to visualize the mitochondria and mitochondrial DNA in wild-type cells.

FLUORESCENCE MICROSCOPY

As yeast cells have been used increasingly for experiments in cell biology, methods for determining the intracellular localization of gene products have been developed, usually by adaptation of methods first used in animal cells. The association of gene products with known cellular structures has frequently led to important insights into the function of the genes involved. Moreover, examination of the morphologies of the nucleus and cytoskeleton gives precise information about the cell cycle stage of individual cells and has been used to stage cell division cycle mutants.

In this experiment, we will use three different methods of identifying structures or proteins in yeast cells using fluorescence microscopy. The first is immunofluorescence, in which fixed cells are incubated first with primary antibodies against the protein of interest, then with fluorescently labeled secondary antibodies directed against the primary antibodies. This layering of primary and secondary antibodies increases the potential fluorescent signal. We use immunofluorescence with anti-tubulin antibodies to visualize the microtubule cytoskeleton. This method has the advantage that it usually generates a strong fluorescence signal, but the disadvantages that cells must be fixed prior to observation and that the cell wall must be removed to allow access of the antibodies. Care must be taken to avoid artifacts due to the cell preparation.

The second fluorescence microscopy method makes use of small molecules that are fluorescent and either bind to specific proteins in the cell or are partitioned to certain organelles. We will use 4′,6-diamidino-2-phenylindole (DAPI) to stain DNA, 3,3′-dihexyloxacarbocyanine iodide (DiOC$_6$) to stain mitochondria, and rhodamine-phalloidin to stain the actin cytoskeleton. DAPI binds specifically to DNA and becomes more intensely fluorescent when bound. DiOC$_6$ is a fluorescent hydrophobic molecule that is specifically transported into mitochondria. Phalloidin is a toxin from the *Amanita phalloides* mushroom that binds to actin filaments; when labeled with rhodamine, it allows the fluorescent visualization of actin filaments. DAPI can be used on fixed or living cells, DiOC$_6$ must be used on living cells, and rhodamine-phalloidin is usually used on fixed cells.

The last and most recently developed method makes use of green fluorescent protein (GFP), a naturally fluorescent protein from the jellyfish *Aequoria victoria*. GFP is remarkable in that it retains its fluorescence when expressed in bacterial, fungal, plant, and animal cells, making it an ideal fluorescent marker protein. To identify the intracellular localization of a protein, the gene for that protein is fused to the GFP coding sequence, such that a fusion protein is created. Expression of this fusion protein in cells allows visualization of the protein in living cells, and (usually) in fixed cells as well. Although care must be taken to ensure that the fusion protein behaves like the wild-type protein, the GFP fusion method is extremely powerful because it allows dynamic behaviors to be observed in living cells. We will look at cells expressing either GFP alone, or a fusion of GFP to *TUB4*, a protein that is localized to the spindle pole body.

STRAINS

1-1	TSY623	*MAT*α *ade2 his3 leu2 ura3*
1-2	TSY807	*MAT*a *his3 leu2 lys2 ura3*
1-3	TSY800	*MAT*a/α *ADE2/ade2 his3/his3 leu2/leu2 lys2/LYS2 ura3/ura3*
1-4	TSY481	*MAT*a/α *ADE2/ade2 his3/his3 leu2/leu2 lys2/lys2 ura3/ura3* [pTS568]
1-5	TSY514	*MAT*a *his3 leu2 lys2 ura3* [pTS592]

PLASMIDS

pTS568	*CEN URA3* GFP
pTS592	*CEN URA3 TUB4*-GFP

PROCEDURE

SAFETY NOTES

Formaldehyde is toxic and is a carcinogen. It is readily absorbed through the skin and is irritating to the eyes, skin, mucous membranes, and upper respiratory tract. Wear gloves and safety glasses and always work in a chemical hood.

DAPI is a possible carcinogen. It may be harmful if it is inhaled, swallowed, or absorbed through the skin. It may also cause irritation. Wear gloves, face mask, and safety glasses, and do not breathe the dust.

Day 1

You will be provided with cultures of strain 1-2 that were fixed during either log phase growth or stationary phase. Examine these using phase or differential interference contrast (DIC) microscopy and count 100 cells of each culture, noting the numbers of unbudded, small budded, and large budded cells.

You will also be provided with a fixed mating mixture of 1-1 and 1-2 cells. Examine this culture using phase or DIC microscopy and identify shmoos and zygotes, based on their distinctive morphology.

Compare the morphologies of haploid 1-2 and diploid 1-3 cells, looking for differences in the size and shape of cells.

Start overnight cultures of:

strain 1-3 in 10 ml of YPD at 30°C

strain 1-3 in 5 ml of YPEG (same as YPD except 3% ethanol and 3% glycerol replace glucose as the carbon source) at 30°C

strains 1-4 and 1-5 in 5 ml of SGal-ura (minimal medium with 2% galactose as carbon source) at 30°C for use the following morning

Day 2

Dilute back the cultures from yesterday if they are overgrown, and grow to early log phase.

(i) Fix 5 ml of the strain 1-3 cells grown in YPD by adding 0.5 ml of formaldehyde directly to culture (total concentration is 3.7%; standard stock solution is 37%). Incubate for 1–2 hours at room temperature (~23°C). Pellet cells in a clinical centrifuge and wash twice with 0.1 M potassium phosphate (pH 7.5). Store cells overnight at 4°C.

(ii) Stain living cells of strain 1-3 grown in YPD and YPEG with $DiOC_6$ according to Techniques & Protocols #10, Yeast Vital Stains.

(iii) Stain living cells of strain 1-3 grown in YPD with Calcofluor according to Techniques & Protocols #10, Yeast Vital Stains.

(iv) Make wet mounts of the strains 1-4 and 1-5 cultures as above and view the fluorescence using a fluorescein filter set.

Day 3

(i) Stain fixed strain 1-3 YPD cells with anti-tubulin primary antibody, FITC-conjugated goat anti-rabbit secondary antibodies, and DAPI, according to Techniques & Protocols #11, Yeast Immunofluorescence.

(ii) Stain fixed 1-3 YPD cells with rhodamine phalloidin, according to Techniques & Protocols #12, Actin Staining in Fixed Cells.

MATERIALS

Day 1   10 ml of YPD
                 5 ml of YPEG
                 10 ml of SGal-ura

Day 2   Formaldehyde (37% stock solution)
                 PBS
                 $DiOC_6$ (10 µg/ml stock solution in ethanol)
                 Calcofluor (1 mg/ml stock solution in 100 mM potassium phosphate [pH 7.5])

Day 3   Zymolyase 100T (120493-1, Seikagaku America Inc.)
                 100 mM Potassium phosphate (pH 7.5)
                 Mercaptoethanol
                 Polylysine
                 PBS
                 Methanol and acetone at –20°C
                 PBS + 3% BSA
                 YOL1/34 anti-tubulin antibody
                 Goat anti-rat fluorescein-conjugated secondary antibody
                 DAPI (Sigma D 9542 or Accurate Chemical and Scientific Corp.)
                 Mounting medium
                 Rhodamine-phalloidin

Isolation and Characterization of Auxotrophic, Temperature-sensitive, and UV-sensitive Mutants

Since spontaneous mutation frequencies are low, yeast is usually treated with such mutagens as ultraviolet (UV) radiation, nitrous acid, ethyl methanesulfonate (EMS), diethyl sulfate, and 1-methyl-nitro-nitrosoguanidine to enhance the frequency of mutants. These mutagens are remarkably efficient and can induce mutations at a rate of 5×10^{-4} to 1×10^{-2}/gene without substantial killing. Even though there are known methods to increase the proportion of mutants by killing off the nonmutants with nystatin and other agents, it is usually unnecessary to use selective means to obtain reasonable yields of mutants (Henry et al. 1975; Snow 1966; Thouvenot and Bourgeois 1971; Walton et al. 1979). In this experiment, auxotrophic, temperature-sensitive, and UV-sensitive mutants will be isolated from EMS-treated yeast.

Auxotrophic mutants have been invaluable for the elucidation of biochemical pathways as well as for the study of the relationship between enzyme structure and function (Lindegren et al. 1965; Lingens and Oltmanns 1964, 1966). Studies on the intermediates accumulated by amino acid auxotrophs have facilitated the unraveling of biochemical pathways.

Studies on temperature-sensitive mutants make it possible to amplify the genetic picture of the genome (Hartwell et al. 1967; Pringle and Hartwell 1981). Many yeast genes specify proteins that participate in indispensable functions (i.e., RNA polymerases, tRNA synthetases, etc.). Mutations that completely destroy the activity of these proteins are lethal, and those that give only partial activity are difficult to work with genetically and biochemically. A mutation that affects the structure of one of these indispensable proteins such that it can function at low temperature but not at high temperature is much more useful. A strain carrying such a mutation grows at normal or nearly normal rates at low temperature but does not grow on any medium at elevated temperatures. This phenotype distinguishes the temperature-sensitive mutation in a vital function from a supplementable temperature-sensitive mutant (i.e., one whose defect is in the biosynthesis of an amino acid). A complete study of temperature-sensitive mutations should allow a description and understanding of the functioning of almost the entire genome. In several bacteriophages, this expectation has been realized.

DNA repair and recombination are amenable to study through the use of UV-sensitive (or Rad⁻) mutants (Cox and Parry 1968; Resnick 1969; Game and Cox 1971; Moustacchi 1972). The logic behind this is that there are enzyme systems that repair UV-induced lesions of DNA. Mutations in the *RAD* genes specifying these enzymes render the cells incapable of repairing UV-induced damage and, therefore, much more sensitive to radiation than wild-type cells. Some of the UV-sensitive mutants also have impaired recombination ability, because some proteins are used for both processes. Because some *RAD* genes control vital functions, their disruption is lethal.

In certain cases, it is valuable to be able to select directly for mutations in particular genes. For example, sometimes it is convenient to introduce auxotrophic markers into strains without having to put them through a genetic cross. Two particularly easy direct selections for auxotrophic mutations are the use of α-aminoadipate (αAA) for selection of *lys2* and *lys5* mutations (Chattoo et al. 1979; Zaret and Sherman 1985) and the use of 5-fluoro-orotic acid (5-FOA) for selection of *ura3* and *ura5* mutations (Boeke et al. 1986). In this exercise we will explore the frequencies and phenotypes of αAA and 5-FOA resistant mutants.

The mutagenesis described here involves treatment of wild-type yeast strains with EMS. Half of the class will mutagenize a *MATa* strain, half a *MATα* strain. After mutagenesis, the strains will be diluted and plated onto complete medium plates at a concentration of about 200 cells/plate. After these cells have grown into colonies, they will be transferred to various media by replica-plating. Temperature-sensitive mutants will be detected by comparing pairs of plates that were incubated at room temperature (about 23°C) and 37°C. UV-sensitive mutants will be detected by comparing irradiated plates with the controls. Auxotrophic mutants will be detected by lack of growth on a minimal medium that contains glucose, potassium phosphate, ammonium sulfate, a few vitamins, salts, and trace metals. The specific requirements can be determined by testing the colony from the original YPD plate on various types of synthetic media.

To determine the specific auxotrophic defects of mutants able to grow on complete but not minimal defined medium, putative mutants will be transferred to nine different minimal (SD) medium plates supplemented with pools of various amino acids, purines and pyrimidines, and other metabolites (see below). From the pattern of growth of a particular strain on these plates, it will be possible to identify the specific requirement of the mutant strain.

Pools	#1	#2	#3	#4	#5
#6	adenine	guanine	cysteine	methionine	uracil
#7	histidine	leucine	isoleucine	valine	lysine
#8	phenylalanine	tyrosine	tryptophan	threonine	proline
#9	glutamate	serine	alanine	aspartate	arginine

Generally, a colony will respond on a plate containing one of the pools from 1 to 5 and on another plate containing one of the pools from 6 to 9, thus allowing direct identification of a single growth factor requirement. For example, a colony growing on pools 1 and 7 requires histidine, whereas a colony growing on pools 3 and 8 requires tryptophan. If a colony grows only on one of the nine pools, it requires more than one of the nutrients in that pool. A mutant blocked early in the pathway of aromatic amino acid biosynthesis will grow only on pool 8.

STRAINS

2-1 S288C   *MAT*α *mal gal2*
2-2 D665-1A *MAT*a

PROCEDURE

SAFETY NOTES

EMS is a potent mutagen. Wear gloves and work in a hood when tubes are open. Neutralize all EMS waste with 5% sodium thiosulfate before discarding. Use disposable tubes and pipettes for all manipulations.

Exposure of the skin or the eyes to UV irradiation is dangerous and should be avoided. To prevent exposure, irradiate plates in an enclosed box equipped with a door.

Day 1

Inoculate 5 ml of YPD with one of the strains above (half of the class will use the *MAT*α strain 2-1, and the remainder of the class will use the *MAT*a strain 2-2). Grow overnight at 30°C.

Day 2

Mutagenize your cells with EMS using the method described in Techniques and Protocols #20, "EMS Mutagenesis." Upon the completion of the mutagenesis protocol, you will have one tube of mutagenized cells and one nonmutagenized control. Three steps will be taken with these cells:

1. The cells will be spread directly on αAA medium and 5-FOA medium to select for *lys2* and *ura3* mutants.

2. The cells will be grown in YPD overnight and then spread on αAA and 5-FOA media.

3. The cells will be diluted and spread on YPD plates for colonies to be screened for other mutations.

First, spread 0.1 ml of the mutagenized cells directly on an αAA plate and on a 5-FOA plate. In addition, as a control, plate 0.1 ml of the washed, nonmutagenized cells onto duplicate αAA and 5-FOA plates. The frequencies of colonies arising from the mutagenized and nonmutagenized cultures will also serve to monitor the frequency of mutagenesis caused by the EMS treatment.

Second, inoculate 0.1 ml of each cell suspension (mutagenized and unmutagenized) into 1 ml of YPD broth. Grow overnight at 30°C.

Third, to achieve the goal of obtaining 200 colonies/YPD plate, dilute your EMS mutagenized cells to a concentration of 2000 cells/ml (based on the hemocytometer count, above, and assuming no loss of cells in the mutagenesis procedure). This should be about a 1:100,000 dilution, but may need to be adjusted depending on the initial concentration of cells used. Spread 0.1, 0.2, and 0.4 ml each on separate YPD plates, using ten plates for each of the three different volumes plated. Incubate all the plates for 4 days at room temperature (23°C).

Day 3

Wash the 1-ml YPD overnight cultures from Day 2 with H_2O and resuspend in 1 ml of H_2O. Spread 0.1 ml onto duplicate αAA and 5-FOA plates. Incubate at 30°C.

Day 6

Examine and count the colonies on your YPD plates, and estimate the survival after EMS. Choose ten or more YPD plates containing approximately 200 colonies/plate for the isolation of (1) auxotrophic mutants, (2) temperature-sensitive mutants, and (3) UV-sensitive mutants. Transfer the colonies from each of the YPD plates by replica-plating onto one SD plate, one SC plate, and four YPD plates. Record the number of colonies transferred. Be sure that each plate is numbered and has an orientation symbol on the back. For detection of auxotrophic mutants, incubate the SC and SD plates for 1 day at 30°C. For temperature-sensitive mutants, incubate one of the YPD replicas per set at 37°C for two days, and one at room temperature (23°C) for two days. For UV-sensitive mutants, we will irradiate one of the YPD replicas per set with UV light (400 setting on a Stratalinker; 40,000 μJoules); this plate and a control, nonirradiated YPD plate, will be incubated 1 day at 30°C. The plates to be irradiated should be refrigerated immediately after the replica has been made. At a set time, these plates will be collected and an instructor will irradiate them.

lys2 *and* **ura3** *Mutants.* Place the αAA and 5-FOA plates from day 2 at 4°C.

Day 7

Auxotrophic Mutants. Compare each of the ten SD plates with the SC plates to identify candidate auxotrophic mutants. There should be 5% such colonies. Try to identify at least 20 candidate colonies. We will use a multipronged inoculating device (a frogger) to transfer cells from each of these colonies onto a series of plates designed to reveal the specific auxotrophic deficiencies of each candidate. To "frog," place 100 μl of sterile H_2O in the wells of a microtiter dish; in each well suspend an aliquot of cells from one of your selected colonies using a sterile toothpick. Take care to add the same amount of inoculum from each colony selected, and to suspend it well. Include also strain 2-1 or 2-2. Transfer a droplet from each well onto the appropriate plates by lowering the prongs of the frogger into the wells, lifting it out quickly, then touching it to the surface of a fresh plate. Using frogging, inoculate the nine pool plates, an SC and an SD plate with cells from your candidate auxotrophs. Incubate overnight at 30°C.

UV-sensitive Mutants. Compare the irradiated plates with the unirradiated controls. Transfer by frogging the colonies not appearing on the irradiated plate onto two YPD plates. Try to test at least ten colonies. Include strain 2-1 or 2-2 also. Irradiate one of the plates as before and incubate the two plates overnight at 30°C.

lys2 *and* ura3 Mutants. Count the number of colonies arising on the αAA and 5-FOA plates spread on Days 2 and 3. Record the data in the Table below. Estimate the frequency of each type of mutant before and after mutagenesis and before and after outgrowth following mutagenesis, taking 2×10^8 cells/ml as the saturation density of your YPD overnight cultures. Purify four colonies from an αAA plate on one YPD plate and four from a 5-FOA plate on another YPD for subsequent analysis of their Lys and Ura phenotypes. Incubate the plates at 30°C. Did the frequency of each type of mutant increase after mutagenesis? Why might growth after mutagenesis affect the recovery of each type of mutant?

Medium:	αAA		5-FOA	
Treatment:	No outgrowth	Outgrowth	No outgrowth	Outgrowth
# of colonies per plate:				

Day 8

Auxotrophic Mutants. Record the growth response of your candidates. Assign each candidate a name that corresponds to its auxotrophy (e.g., leu#1, leu#2, etc.). Record the names of your auxotrophs in the appropriate categories of the classwide listing of auxotrophs. Also label the corresponding auxotrophs on your SD plates from Day 7. You will be assigned a specific category of auxotrophs with which to perfrom a complementation test. Collect all the auxotrophs of that category from your classmates. Make two streak master YPD plates of your category of auxotrophs: one for the MAT**a** strains and one for the MATα strains. Include the 2-1 on the MATα plate and 2-2 on the

MATa plate. Each streak plate should have up to eleven parallel streaks that continue across the entire surface of the plate (see Appendix D for a template). The top and bottom streaks should be about 2 cm from the top and bottom of the plate (respectively). Incubate overnight at 30°C then refrigerate.

UV-sensitive Mutants. Record the growth response and assign a number to each mutant. Make two identical streak master YPD plates with up to ten of your mutants (plus the 2-1 or 2-2 control). Incubate overnight at 30°C, then refrigerate.

Temperature-sensitive Mutants. Compare the 37°C plate with the 23°C plate. Identify up to 20 colonies that failed to grow at the high temperature. Use the frogging method to inoculate cells from these Ts⁻ candidates, as well as either the 2-1 or 2-2 control, onto two YPD plates. Also frog onto YPD containing 30% sucrose. This plate will allow us to identify those temperature-sensitive mutants whose defects cause osmotic lysis on the YPD plate or whose defects are suppressed by high osmotic conditions (osmotic-remedial mutants). Incubate one YPD plate and the YPD + sucrose plate at 37°C, and the other YPD plate at room temperature for two days.

lys2 and ura3 Mutants. Frog the purified αAAᴿ colonies onto SD and SD + lys plates to test for the Lys phenotypes. Frog the purified 5-FOAᴿ colonies onto SD and SD + ura plates to test the Ura phenotype.

Day 10

Auxotrophic Mutants. Sequentially replica-plate the MATa and MATα streak plates onto one velvet such that the streaks are perpendicular. Print from this velvet onto a fresh YPD plate. Incubate at 30°C.

UV-sensitive Mutants. Exchange one of your streak plates for a streak plate of UV sensitive mutants of the opposite mating type. Use replica-plating to create a cross-streaked YPD plate. Incubate overnight at 30°C.

Temperature-sensitive Mutants. Record the growth of your mutants at 23°C and 37°C. Assign a number to each mutant. Make two identical streak masters of your mutants (plus the 2-1 or 2-2 control) on a pair of YPD plates. Incubate two days at room temperature.

lys2 and ura3 Mutants. Record the growth of the lys2 and ura3 mutants from Day 9.

Day 11

Auxotrophic Mutants. Replica-plate the cross-streaked mutants to one SC and one SD plate. Incubate at 30°C.

UV-sensitive Mutants. Replica-plate your cross-streaked mutants to two YPD plates, then refrigerate immediately. At a prescribed time, a lab assistant will collect your plates, irradiate one plate of each pair ,and incubate both plates overnight at 30°C. A diploid is expected to have normal UV sensitivity if it was formed from two recessive Rad⁻ strains with mutations at different loci.

Day 12

Auxotrophic Mutants. Score your complementation test. How many cases of non-complementation did you observe? Explore *saccharomyces* databases to estimate the number of genes that can be mutated to yield the auxotrophic phenotype you are studying.

UV-sensitive Mutants. Score the complementation tests.

Temperature-sensitive Mutants. Exchange one of your streak plates for a streak plate of Ts⁻ mutants of the opposite mating type. Use replica-plating to create a cross-stamped YPD plate; incubate at 23°C for two days.

Day 14

Temperature-sensitive Mutants. Replica-plate your cross-streaked mutants to 2 YPD plates; incubate one at 37°C and the other at 23°C for two days.

Day 16

Temperature-sensitive Mutants. Score the complementation tests.

MATERIALS

NOTE: Amounts provided are the requirements for each pair.

Day 1	1 culture tube, containing 5 ml of YPD
Day 2	10 ml of sterile distilled H_2O
	5 ml of sterile 0.1 M sodium phosphate buffer (pH 7)
	EMS (methanesulfonic acid ethyl ester; Sigma M 0880)
	1 ml sterile 5% sodium thiosulfate (w/v)
	2 culture tubes, each containing 1 ml of YPD
	2 αAA plates
	2 5-FOA plates
	30 YPD plates
Day 3	10 ml sterile H_2O
	2 αAA plates
	2 5-FOA plates
Day 6	10 SC plates
	10 SD plates
	40 YPD plates
	10 sterile velveteen squares

Day 7   4 YPD plates
1 each of the nine pool plates
1 SD plate
1 SC plate
10 ml of sterile H_2O
2 sterile microtiter dishes
95% ethanol (for flaming frogger)

Day 8   6 YPD plates
2 SD plates
1 SD + lys plate
1 SD + ura plate
1 YPD plate containing 30% sucrose
1 sterile microtiter dish
10 ml of sterile H_2O
95% ethanol (for flaming frogger)

Day 10   4 YPD plates
2 sterile velveteen squares

Day 11   2 YPD plates
1 SD plates
1 SC plate
2 sterile velveteen squares

Day 12   1 YPD plate
1 sterile velveteen square

Day 14   2 YPD plates
1 sterile velveteen square

REFERENCES

Auxotrophic Mutants

Lindegren G., Hwang, L.Y. Oshima Y., and Lindegren C. 1965. Genetical mutants induced by ethyl methanesulfonate in *Saccharomyces*. *Can. J. Genet. Cytol.* **7**: 491–499.

Lingens F. and Oltmanns O. 1964. Erzeugung und untersuchung Biochemischer and Mangelmutanten von *Saccharomyces cerevisiae*. *Z. Naturforsch.* **19B**: 1058–1065.

——. 1966. Uber die Mutagene Wirkung von 1-nitroso-3-nitro-1-methyl-guanidin (NNMG) und *Saccharomyces cerevisiae*. *Z. Naturforsch.* **21B**: 660–663.

Temperature-sensitive Mutants

Hartwell L.H. 1967. Macromolecule synthesis in temperature-sensitive mutants of yeast. *J. Bacteriol.* **93**: 1662–1670.

Pringle J.R. and Hartwell L.H. 1981. The *Saccharomyces cerevisiae* cell cycle. In *The molecular biology of the yeast* Saccharomyces: *Life cycle and inheritance* (ed. J.N. Strathern et al.), pp. 97–142. Cold Spring Harbor Laboratory, Cold Spring Harbor, New York.

UV-sensitive Mutants

Cox B.S. and Parry J.M. 1968. The isolation, genetics and survival characteristics of ultraviolet-light-sensitive mutants in yeast. *Mutat. Res.* **6**: 37–55.

Game J.C. and Cox B.S. 1971. Allelism tests of mutants affecting sensitivity to radiation in yeast and a proposed nomenclature. *Mutat. Res.* **12**: 328–331.

Moustacchi E. 1972. Evidence for nucleus independent steps in control of repair of mitochondrial damage. I. UV-induction of the cytoplasmic "petite" mutation in UV-sensitive nuclear mutants of *Saccharomyces cerevisiae*. *Mol. Gen. Genet.* **114**: 50–58.

Resnick M.A. 1969. Genetic control of radiation sensitivity in *Saccharomyces cerevisiae*. *Genetics* **62**: 519–531.

Selections for Auxotrophs

Boeke J.D., LaCroute F., and Fink G.R.. 1986. A positive selection for mutants lacking orotidine-5'-phosphate decarboxylase activity in yeast; 5'-fluoro-orotic acid resistance. *Mol. Gen. Genet.* **197**: 345–346.

Chattoo B.B., Sherman, F. Azubalis D.A., Fjellstedt T.A., Mehnert D., and Ogur M. 1979. Selection of *lys2* mutants of the yeast *Saccharomyces cerevisiac* by the utilization of α-aminoadipate. *Genetics* **93**: 51–65.

Zaret K.S. and Sherman F. 1985. α-Aminoadipate as a primary nitrogen source for *Saccharomyces cerevisiae* mutants of yeast. *J. Bacteriol.* **162**: 579–583.

Enrichment Methods

Henry S.A., Donahue T.F., and Culbertson M.R. 1975. Selection of spontaneous mutants by inositol starvation in *Saccharomyces cerevisiae*. *Mol. Gen. Genet.* **143**: 5–11.

Snow R. 1966. An enrichment method for auxotrophic yeast mutants using the antibiotic "nystatin." *Nature* **211**: 206–207.

Thouvenot D.R. and Bourgeois C.M.. 1971. Optimisation de la selection de mutants de *Saccharomyces cerevisiae* par la nystatine. *Ann. Inst. Pasteur* **120**: 617–625.

Walton B.F., Carter B.L.A., and Pringle J.R. 1979. An enrichment method for temperature-sensitive and auxotrophic mutants of yeast. *Mol. Gen. Genet.* **171**: 111–114.

Meiotic Mapping

The events of meiosis make it possible to deduce information about the positional relationships of genetic markers. For many years, meiotic mapping has been of great utility in constructing genetic maps (Mortimer and Hawthorne 1966). Fincham et al. (1979) and Mortimer and Hawthorne (1969) give general information concerning meiotic mapping and tetrad analysis. In present times, genetic mapping techniques are instrumental in monitoring the manipulations of the genome that are performed using molecular genetic techniques. The four spores in an ascus are the products of a single meiotic event, and the genetic analysis of these tetrads can provide linkage relationships of genes present in the heterozygous condition. It is also possible to map a gene relative to its centromere if known centromere-linked genes are present in the hybrid. Although the isolation of four spores from an ascus is a skill acquired only with considerable practice, tetrad analysis is useful not only for linkage studies, but also for constructing strains necessary for genetic and biochemical experiments.

In a cross of two haploids, $AB \times ab$, followed by tetrad dissection, the segregation of the markers can yield three types of tetrads. The three classes of tetrads—parental ditype (PD), nonparental ditype (NPD), and tetratype (T)—from a diploid that is heterozygous for two markers, $AB \times ab$, are shown in the following table:

	PD	NPD	T
	AB	aB	AB
	AB	aB	Ab
	ab	Ab	ab
	ab	Ab	aB
Random assortment:	1 :	1 :	4
Linkage:	>1 :	<1	
Centromere linkage:	1 :	1 :	<4

LINKED MARKERS

The ratio of the three types of tetrads that is observed for a pair of markers is a function of the relative map positions of the markers. If the markers A and B segregate at random with respect to one another, the PD, NPD, and T patterns will be observed in a 1:1:4 ratio. However, if they are linked they will yield different ratios that can be used to deduce their map distances. Figure 1 shows the outcomes of zero, one, or two crossovers between markers A and B on the same chromosome. When there are no crossovers, a parental ditype is always formed. A single crossover always yields a tetratype. There are four possible types of double crossovers involving two, three, or four strands (chromatids). Two-strand double crossovers yield a PD, the two types of three-strand double crossovers both yield tetratypes, and a four-strand double crossover yields an NPD. Meioses in which more than two crossovers occur between A and B can result in any of the three types of tetrads, depending on which strands are involved. The probability that a crossover will occur between two markers is approximately proportional to the physical distance between them. Thus, markers that are closely linked yield predominantly PD tetrads and few NPD tetrads. In contrast, markers that are unlinked yield equal numbers of PD and NPD tetrads.

The distinctive segregation patterns exhibited for markers between which there have been zero, one, or two crossovers allow the formulation of a mapping function (Perkins 1949). For genetic intervals that experience at most two crossovers, the NPDs provide an accurate indicator of the frequency of double crossovers, which in turn allows us to deduce the incidence of single crossovers. The mapping function is an indicator of the average number of meiotic crossovers per chromatid between markers A and B. With a single crossover (SCO), half of the chromatids in the tetrad become recombinant, whereas double crossovers result in an average of one crossover per chromatid (although the crossovers do not always occur on all four chromatids; Fig. 1). The mapping function can therefore be expressed as:

$$100 \times \frac{1/2(\text{SCO tetrads}) + (\text{DCO tetrads})}{\text{Total tetrads}}$$

This function gives us the number of centiMorgans (cM) between our markers. To estimate the total number of tetrads in which a DCO occurred in our interval, we take advantage of the fact that one fourth of the DCO tetrads yield NPD tetrads (Fig. 1).

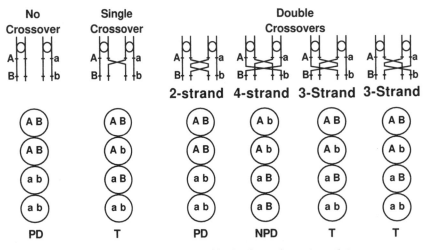

Figure 1. Segregation of linked markers A and B.

With this information we can estimate the total number of DCO tetrads, assuming no chromatid interference, as 4(NPD tetrads). SCO tetrads are always tetratypes (Fig. 1), but tetratypes are also generated by DCO tetrads. To estimate the number of SCO tetrads, we subtract the contribution of DCO tetrads to the pool of tetratype tetrads. Because DCOs yield two Ts for each NPD (Fig. 1), this is 2(NPD). Placing these values in the equation above gives us the following mapping function:

$$100 \times \frac{1/2\,(T - 2[NPD]) + 4(NPD)}{\text{Total tetrads}} = 100 \times \frac{1/2T + 3(NPD)}{\text{Total tetrads}}$$

UNLINKED MARKERS

A number of assumptions concerning interference must be made to determine map distances between larger intervals. In addition, for larger intervals, we cannot tell whether an NPD tetrad is the consequence of two, three or even more crossovers between A and B. The mapping function, above, is based on the assumption that there are at most two crossovers in the interval per meiosis and therefore results in underestimates of the true genetic distance for intervals that experience multiple crossovers per meiosis. The only accurate way of measuring long intervals is by the summation of shorter intervals.

A special case of unlinked markers exists for those that are on different chromosomes but are close to their respective centromeres. Such situations yield equal numbers of PD and NPD tetrads (remember, PD=NPD is the hallmark of unlinked markers), but in this case instead of observing the PD:NPD:T ratio of 1:1:4, characteristic of random assortment, there is a reduction in the proportion of T asci.

The distance from a gene to its centromere can be estimated by determining the frequency of second-division segregation (SDS), explained below. First-division segregation is the term applied to the movement of homologous centromeres to opposite poles at meiosis I (Fig. 2A). Heterozygous genes that map adjacent to the centromeres, such as *TRP1/trp1* (Fig. 2A), also show this segregation pattern in most meioses. Only when there is a crossover in the small interval between the *TRP1* locus and the nearby centromere do the *TRP1* and *trp1* alleles migrate away from each other at meiosis II, exhibiting SDS (Fig. 2B). For *TRP1* this occurs in only about 1% of meioses, making the segregation of *TRP1* a good indicator of centromere behavior in meiosis. It is possible to determine the frequency of SDS for a gene of interest by comparing its segregation to the segregation of *TRP1*. If we perform tetrad analysis with a strain heterozygous for our gene of interest and *TRP1* (*A, TRP1* × *a, trp1*) then PD and NPD tetrads will indicate instances of FDS of gene A (Fig. 3A), and T tetrads will indicate SDS (Fig. 3B). The map distance between gene A and its centromere can be approximated as:

$$100 \times \frac{\text{(Tetratype tetrads)}}{2\,\text{(Total tetrads)}}$$

This mapping function is based on the assumptions that (1) T tetrads are the result of a single crossover between the gene of interest and its centromere and (2) the PD and

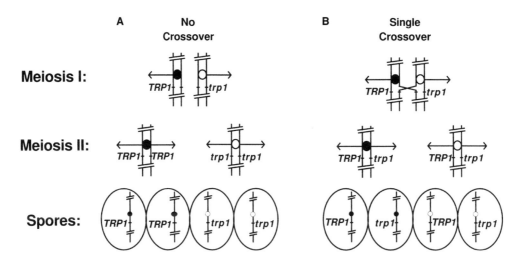

Figure 2. First- and second-division segregation. The gene, *TRP1* is located close to the centromere of chromosome *IV*. In most meioses there are no crossovers between *TRP1* and *CEN4*. In meioses with no crossovers in the *TRP1* to *CEN4* interval (*A*), if a strain is heterozygous for *TRP1* (*TRP1* and *trp1*), the alleles migrate away from each at the first meiotic division, as shown above. This is called first-division segregation. In a fraction of meioses there will be a crossover between the *TRP1* locus and *CEN4* as shown above (*B*). The result is that each homolog carries a TRP1 allele on one chromatid and a trp1 allele on the other. In this situation, the TRP1 and trp1 alleles do not migrate away from each other until meiosis II. This is called second-division segregation.

NPD tetrads have experienced no crossovers in this interval. Therefore, this mapping function severely underestimates the true map distance for intervals that experience multiple crossovers.

It is also possible to determine the percentage of SDS with reasonable accuracy if the hybrid contains two or more centromere-linked markers that may not be as close as, for example, *trp1*. In this case, the SDS array is decided upon by the best agreement among the centromere-linked markers.

Some of the key points to remember when analyzing tetrad data, and a decision tree to aid in drawing conclusions about genetic map relationships of the genes being analyzed, are shown in Figure 4.

AN INTRODUCTION TO TETRAD DISSECTION

Meiosis and sporulation are usually complete within 4 days following the introduction of diploid cells to sporulation medium. A sporulated culture contains a mixture of unsporulated diploid cells, of asci with four haploid spores, and of asci with fewer than four spores. The spores will not germinate or divide on sporulation medium. Most sporulated cultures may be stored in the refrigerator for several months with only a gradual loss of viability. The first step in dissection consists of treating the sporulated

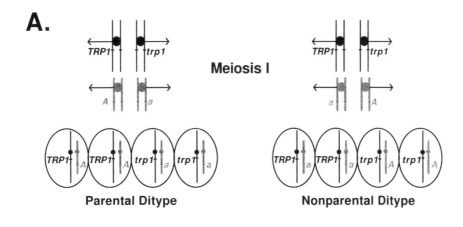

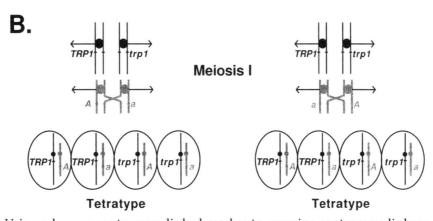

Figure 3. Using a known centromere-linked marker to examine centromere linkage of another gene. (*A*) In the cross *A TRP1* x *a trp1*, when no crossovers occur between our gene of interest (*A* in this example) and its centromere, gene *A* will show two types of segregation with respect to *TRP1*. The two outcomes are determined by the orientations with which the corresponding centromeres attach to the metaphase spindle. These meioses yield PD and NPD tetrads with equal likelihood. (*B*) Regardless of the orientations of the respective centromeres as they attach to the metaphase spindle, if there has been a single crossover between A and its centromere, what would have been a PD or NPD tetrad is converted into a T tetrad.

culture with Zymolyase, which digests the ascus wall but does not disturb the association of the four spores from the ascus. Digested asci are carefully transferred onto a YPD plate using a sterile inoculating loop. The four spores are then separated from each other and moved to isolated positions on the YPD plate using a microneedle attached to a micromanipulator. The spores will germinate and form colonies in 2–3 days after dissection. The spore clones can be transferred to slants or to YPD plates for storage. A master plate is prepared by inoculating a YPD plate with the strains; most of the phenotypes are scored by replica-plating onto appropriate media. The mating types, complementarity of markers with similar phenotypes, and the nature of mutant alleles within a single gene can be determined by replica-plating the master plate onto two

A. Key Points

1) Phenotypes caused by mutations in a single gene segregate 2:2.

2) For unlinked genes, parental ditype (PD) = nonparental ditype (NPD).

3) For unlinked genes, at least one of which is not linked to a centromere, PD:NPD:T = 1:1:4

4) For unlinked genes that are both linked to a centromere, tetratypes (T) result from crossing-over between the genes and the centromeres.

6) For linked genes, PD > NPD

7) The map distance of small genetic intervals that experience at most two crossovers per meiosis can be extimated as:

$$\text{centimorgans (cM)} = 100 \times \frac{1/2\ T + 3\ NPD}{\text{Total}}$$

B. Decision Tree

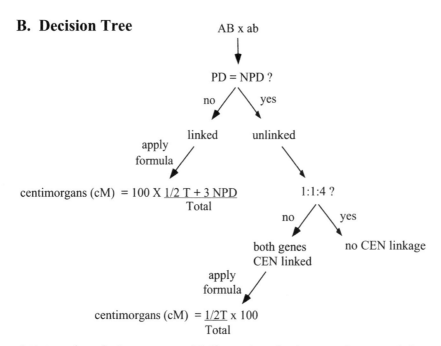

Figure 4. A tetrad analysis summary. (*A*) Key points for interpreting tetrad data. (*B*) A decision tree to aid in evaluating tetrad data.

YPD plates and replica-plating onto these prints with lawns of *MAT***a** and *MAT*α tester strains with the appropriate genotypes. The mating types of the spore colonies are revealed by their ability to form diploids when replica-plated onto *MAT***a** and *MAT*α lawns. Because we know the exact genotypes of the test lawns, the phenotypes of the diploids formed between the lawns and the spore colonies allow us to deduce the genotypes of the spore colonies. A description of a variety of methods and instruments used for tetrad analysis can be found in Sherman and Hicks (1991).

Many of the techniques and principles of chromosome mapping will be illustrated by tetrad analysis of the diploid 3-3, which was constructed by a cross of strains 3-1 and 3-2. A sporulated culture of the above strain will be furnished. Since micromanipulation requires considerable practice, the schedule will commence after a number of asci have been successfully dissected.

STRAINS

3-1	DDY700	*MAT**a**, ura3-52, leu2-3,112, arg4–ΔBglII, trp2, cyh2*
3-2	DDY701	*MAT*α, *ura3-52, trp1-289, arg4-Δ42, ade1*
3-3	610.6D x 611.10D	
3-4	AAY1018	*MAT**a**, his1* (same strain as 9-3)
3-5	AAY1017	*MAT*α, *his1* (same strain as 9-2)
3-6	DDY702	*MAT**a**, arg4-Δ42, his3Δ1, trp1-289, ade2,*
3-7	DDY703	*MAT*α, *arg4-Δ42, his3Δ1, trp1-289*
3-8	DDY704	*MAT**a**, arg4-ΔBglII, his3Δ1, trp1-289*
3-9	DDY705	*MAT*α, *arg4-ΔBglII, his3Δ1, trp1-289*
3-10	NE29	*MAT**a**, trp1-289, ura3-52*
3-11	NE30	*MAT*α, *trp1-289, ura3-52*
3-12	DRM117.39C	*MAT**a**, trp2, leu2-3,112, ura3-52, arg4ΔHpa, rad3*
3-13	DRM117.126A	*MAT*α, *trp2, leu2-3,112, ura3-52, arg4ΔHpa, ilv1-92*

Note: Strain 3-3 may sporulate after long-term storage on YPD. The diploid may be reconstructed by mating strains 3-1 and 3-2.

PROCEDURE

Day 1

Preparation of the Microneedle. A method for constructing a tetrad dissection needle will be demonstrated in class (Scott and Snow 1978). Use the demonstrated approach, which is described in Method 1 of Techniques and Protocols #22 (Making a Tetrad Disssection Needle), to construct your own needle.

Treatment with Zymolyase. Your instructors introduced diploid cells to sporulation medium 3 days prior to the beginning of the course. Examine the sporulated culture under a microscope and identify the unsporulated cells, the four-spored asci, and the asci with fewer than four spores. Each student in the class will be allotted 3 days to produce twenty dissected tetrads with four viable spores. The methods used to treat cells with zymolyase, prepare a dissection plate, and micromanipulate tetrads are described in Techniques and Protocols #21 (Tetrad Dissection).

Day 3

Complete dissection of tetrads.

Day 5

Using the *thin ends* of sterile toothpicks and following the grid provided (see Appendix D), each student should prepare two master plates with spore colonies that have four viable members (ten tetrads per plate). Include the two parental control strains on the bottom of each plate. Incubate overnight at 30°C.

Day 6

Replica-plating Tetrads to Test Media. Replica-plate the master plate to these media: YPD, SC-trp, SC-leu, SC-arg, SC-ade, YPD+cyh, and six YPD plates (see hints below before you begin). The first YPD serves as a fresh master plate for future use. The remaining YPD plates are for matings of your spore colonies with tester strains. Incubate all plates at 30°C.

Hints: twelve replicas is a large number to make from one master plate. After the first six replicas, put a fresh velvet on the replica-plating block, and make a fresh print with your master plate. Because you are using your master plate twice, be careful to make a light print on the first velvet so that there will be cells remaining on the master plate for the second print. Look at each replica plate when you produce it. You should be able to see a light deposition of cells corresponding to the pattern on your master plate. If you don't, try again. (As an alternative method, sterile H₂O in a microtiter dish can be inoculated with spore colonies and the cells can be transferred to the different test plates using the multiprong inoculator [frogging technique]. This will be demonstrated by the instructors.)

Preparation of Mating-type and Allele Testers. Inoculate 5 ml of YPD with the tester strains 3-4 through 3-13. Incubate at 30°C.

Day 7

Preparation of Mating-type Test Lawns. To prepare mating-type test lawns, centrifuge the 5-ml cultures of 3-4 and 3-5 (5 minutes at 2000 rpm in a tabletop centrifuge). Resuspend in 4 ml of YPD in a sterile screwtop tube. Add 0.15 ml of sterile H₂O and 0.15 ml of the resuspended 3-4 culture to 2 SD plates (store the remaining cells at 4°C). Distribute the inoculum evenly with a sterile rod, place the plate on a level surface, and allow the plate to dry. Repeat with the 3-5 cells.

Preparation of Allele and Complementation Test Lawns. These lawns will be prepared using the same approach as is used for mating-type lawns except that both the *MAT*a and *MAT*α versions of the test strain will be combined on the same plate. Pellet and resuspend the 5-ml cultures of strains 3-6 through 3-13. You will prepare four types of test lawns (*arg4-Δ42, arg4-ΔBglII, trp1-289,* and *trp2*), and need two plates for each. For the *arg4* test lawns, inoculate two SC-ura-his plates with 0.15 ml of both 3-6 and 3-7, or 0.15 ml of both 3-8 and 3-9. For the Trp complementation lawns, inoculate two SC-trp plates with 0.15 ml of both 3-10 and 3-11, or 0.15 mls of both 3-12 and 3-13. Distribute the inocula evenly with a sterile rod, place the plates on a level surface, and allow them to dry.

Replica-plating to Test Lawns. Use the six YPD replicas of your tetrads that you made yesterday as the source of cells for printing onto your six test lawns (after they have dried). For each test lawn, place a fresh velvet on the replica-plating block, make a print on the velvet with one of your YPD plates, then place the test lawn plate on the velvet to transfer the tetrads onto the test lawn. Incubate the test plates overnight at 30°C.

Score growth on SC-trp, SC-leu, SC-arg, SC-his, SC-ade, and YPD+cyh. Record the growth phenotypes on the tetrad scoring sheet (Appendix H). Use your master plate as a source of cells for retesting ambiguous spore colonies.

Day 9

Replica-plate *arg4* allele-testing matings to SC-arg and place in the refrigerator. At the end of the day these plates will be collected by your lab assistant, who will irradiate them with 7500 μJoules of ultraviolet light (254 nm) using a Stratagene Stratalinker (set on 75).

Score growth of mating-type test plates and Trp complementation test plates.

Day 10/11

Score growth of *arg4* allele-testing plates.

Day 11

Determine the number of PD, NPD, and T tetrads for each pairwise combination of markers segregating in your cross. Record the PD/NPD/T data on the scoring sheet at the end of this section. Also record the frequency of second-division segregation for each marker. Determine the distance between each gene and its centromere by searching for map information about each gene on the Saccharomyces Genome Database. Use the data you have gathered to calculate:

1. The map distance of each gene scored from all other genes scored.

2. The map distance between each gene and its centromere (How does this compare to the information provided by the database?).

3. The gene conversion frequencies (3:1 and 1:3 segregation) for all alleles we studied.

Day 14

Be prepared to present your data.

MATERIALS

Note: Amounts provided are for 20 tetrads—each group requires double these quantities.

Day 1   Capillary pipettes for preparing microneedle holders
Super Glue
Fiber-optic glass
Zymolyase 100T (120493-1, Seikagaku America Inc.) solution
(0.5 mg/ml in 1 M sorbitol)
Sterile distilled H_2O
4 YPD plates

Day 5   2 YPD plates

Day 6   2 Plates each of:
SC-ade
SC-arg
SC-leu
SC-trp
YPD+cyh
14 YPD plates
4 sterile velveteen pads
10 culture tubes, each containing 5 ml of YPD

Day 7   10 culture tubes, each containing 4 ml of YPD
5 ml of sterile H_2O
4 SD plates
4 SC-his-ura plates
4 SC-trp plates
12 sterile velveteen pads

Day 9   4 SC-arg plates

REFERENCES

Fincham J.R.S., Day P.R., and Radford A. 1979. *Fungal genetics.* University of California Press, Berkeley and Los Angeles.

Mortimer R.K. and Hawthorne D.C. 1966. Genetic mapping in *Saccharomyces. Genetics* **53:** 165–173.

——. 1969. Yeast genetics. In *The yeasts* (ed. A.H. Rose and J.S. Harrison), vol. 1, pp. 385–460. Academic Press, New York.

Mortimer R.K. and D. Schild. 1981. Genetic mapping in *Saccharomyces cerevisiae.* In *The molecular biology of the yeast* Saccharomyces: *Life cycle and inheritance* (ed. J.N. Strathern et al.), pp. 11–26. Cold

Spring Harbor Laboratory, Cold Spring Harbor, New York.

Perkins D.D. 1949. Biochemical mutants in the smut fungus *Ustilago maydis*. *Genetics* **34**: 607–626.

Scott K.E. and Snow R. 1978. A rapid method for making glass micromanipulator needles for use with microbial cells. *J. Gen. Appl. Microbiol.* **24**: 295–296.

Sherman F. and Hicks J. 1991. Micromanipulation and dissection of asci. *Methods Enzymol.* **194**: 21–37.

Experiment III: Tetrad analysis; 3-3 Group no._____

Names_____

		MAT	TRP1	TRP2	CYH2	ARG4	LEU2	ADE1
MAT	P							
	N							
	T							
TRP1	P							
	N							
	T							
TRP2	P							
	N							
	T							
CYH2	P							
	N							
	T							
ARG4	P							
	N							
	T							
LEU2	P							
	N							
	T							

	MAT	TRP1	TRP2	CYH2	ARG4	LEU2	ADE1
%SDS (with TRP1)							
Gene/ CEN Distance (SGD)							

Mitotic Recombination and Random Spore Analysis

In diploid strains, alterations leading to loss of genetic information can be generated by mitotic crossing over, gene conversion, or chromosome loss. Because such events are usually observed as the loss of one of two alleles in a heterozygous strain, they are referred to as loss of heterozygosity (LOH) events. Mitotic crossing over and gene conversion are illustrated in Figure 1. Mitotic crossing over results in LOH of all markers located distal to the point of exchange on the chromosome arm. Mitotic gene conversion results in nonreciprocal exchange of only a small chromosomal region (i.e., a single gene or closely linked genes) and is believed to be analogous to the irregular segregations that are observed at low frequency after meiosis (meiotic gene conversion). Chromosome loss, which results in LOH of markers on both arms of a chromosome, produces a 2N-1 diploid that may or may not have impaired growth, depending on the particular chromosome.

Mitotic crossing over between two nonsister chromatids is likely to occur during the G2 stage of the cell cycle (see Fig. 1). After mitosis, a crossover event can lead to two daughter cells that are homozygous for part of the chromosome arm that was heterozygous in the mother cell (this will occur in half of the recombination events because of the random orientation of chromosomes on the mitotic spindle). Thus, mitotic recombination can lead to the "uncovering" of recessive markers. For some markers, "papillae" or "sectors" derived from the homozygous daughter cells can be observed in a background of heterozygous cells, resulting in papillating or sectoring colonies. As a result of mitotic crossing over and subsequent mitotic segregation, all markers distal to the site of recombination on the chromosome arm are simultaneously homozygosed. It is possible to generate a recombinational map of the chromosomal arm using mitotic recombination data. There is an approximately linear relationship between the frequency of homozygosis of a gene and its distance from the centromere. Thus, quantitation of the relative frequencies of different classes of recombinants can be used to determine the relative distances between two markers and between a marker and its centromere.

Mitotic gene conversion also results from recombination between two nonsister chromatids in G2 (see Fig. 1). In this case, only a limited chromosomal region is homozygosed, with the result that distal markers retain their heterozygosity. If two or

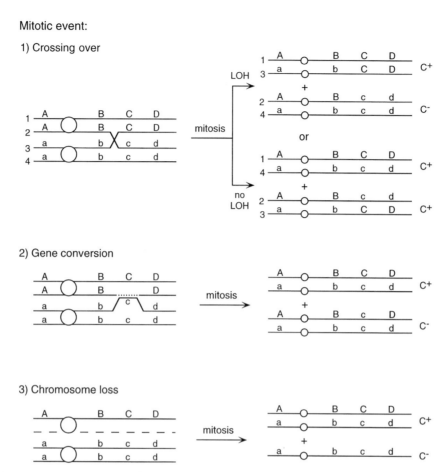

Figure 1. Mitotic segregation of heterozygous markers in a diploid strain after either (1) crossing over; (2) mitotic gene conversion; or (3) chromosome loss.

more markers sector together, this event most likely is due to mitotic crossing over and not to gene conversion, because conversion tracts are generally short (several kilobases at the most).

It is important to note that the spontaneous rate of mitotic recombination is low, ranging from ~10^{-6} to ~10^{-4}, depending on the distance of a particular gene from its centromere and other less well-defined factors. Chromosome loss events occur at similarly low frequencies. To find rare LOH events easily, it is usually necessary to apply a selection or screen. In this experiment, we will use selections for canavanine resistance and 5-fluoro-orotic acid (5-FOA) resistance to identify cells in which a LOH event has occurred.

EXPERIMENTAL DESIGN

A diploid will be created by mating the following two haploid strains:

4-1 TSY812   MATα *can1 hom3 leu2 lys2 ura3*
4-2 TSY813   *MAT**a** ade2 his1 lys2 trp1*

URA3 and *CAN1* are on the left arm of chromosome V, and *HIS1* and *HOM3* are on the right arm of chromosome V. The *URA3* and *CAN1* genes are particularly useful for this experiment because they have the unusual feature that strong selection techniques exist for <u>recessive</u> alleles of both genes. Starting with a diploid that is heterozygous at both loci, *ura3* strains that arise by mitotic recombination, chromosome loss, or meiosis can be selected by their resistance to 5-FOA, and *can1* strains by their resistance to canavanine.

The *URA3* gene encodes orotidine-5′-phosphate decarboxylase, an enzyme that catalyzes one of the steps in pyrimidine synthesis. 5-FOA is taken up by cells and is converted to the toxic compound 5-fluorouracil by the action of the decarboxylase. *URA3/URA3* homozygotes, *URA3/ura3* heterozygotes, and *URA3*/hemizygotes (one copy of a gene in a diploid strain) are all able to convert 5-FOA to 5-fluorouracil, and are sensitive to 5-FOA (5-FOA^S), whereas *ura3/ura3* homozygotes and *ura3*/hemizygotes lack the decarboxylase activity and are resistant to 5-FOA (5-FOA^R).

The *CAN1* gene encodes the arginine permease, which allows uptake of arginine from the medium. Canavanine is a toxic analog of arginine that is taken up by cells through the arginine permease. *CAN1/CAN1* homozygotes, *CAN1/can1* heterozygotes, and *CAN1*/hemizygotes are all able to take up canavanine, and so are sensitive to canavanine (Can^S), whereas *can1/can1* homozygotes and *can1*/hemizygotes lack the permease and are resistant to canavanine (Can^R).

The frequency of LOH at each locus will be found by determining the frequency of Can^R and 5-FOA^R mitotic segregants that arise from the diploid strain 4-1 × 4-2. These frequencies will be compared to the frequency of spontaneous mutations in *CAN1* and *URA3* in the haploid strain 4-2. Mitotic recombination will be distinguished from chromosome loss by examining the Hom3 phenotype on the opposite arm of chromosome V. The relative positions of *CAN1* and *URA3* with respect to the centromere on chromosome V will be determined by seeing how often LOH occurs at *CAN1* but not at *URA3*, or at *URA3* but not at *CAN1*.

An additional part of this experiment will be to examine meiotic recombination in the 4-1 × 4-2 diploid using random spore analysis. This method allows meiotic products to be isolated and scored without performing tetrad analysis. It is particularly useful when many crosses must be analyzed or a rare recombinant must be identified. It is important to realize, however, that information about centromere linkage and the ability to use the tetrad mapping function for more accurate determination of genetic distance are lost when random spore analysis is used in place of tetrad analysis. The principle behind the method is that haploid meiotic segregants are selected away from unsporulated diploid cells by selecting for canavanine resistance in a strain that was originally a *CAN1/can1* heterozygote. Half of all meiotic segregants will be Can^R because they will have received the *can1* gene.

STRAINS

4-1 TSY812   *MATα can1 hom3 leu2 lys2 ura3*
4-2 TSY813   *MAT**a** ade2 his1 lys2 trp1*

PROCEDURE

Day 1

Subclone strains 4-1 and 4-2 to a YPD plate. Incubate at 30°C.

Day 3

In the morning, mate strains 4-1 and 4-2. Transfer a single colony of 4-1 and a single colony of 4-2 to a YPD plate and mix together in a patch (a drop of sterile H_2O will facilitate mixing). Incubate at least 5 hours at 30°C.

In the afternoon, streak out cells from the mixed patch to an SD+lys plate to select for diploids. Incubate at 30°C.

Day 5

(i)   *LOH.* In the evening, inoculate 5 ml of YPD with the 4-1 x 4-2 diploid and 5 ml of YPD with the 4-2 haploid as a control. Incubate on the roller drum at 30°C.

(ii) *Meiotic recombination.* To prepare the 4-1 x 4-2 diploid for sporulation, patch onto a YPD plate and incubate overnight at 30°C.

Day 6

(i)   *Mutation and LOH.* Prepare tenfold serial dilutions of the saturated culture of the 4-1 x 4-2 diploid in sterile microfuge tubes. To determine the frequency of LOH, plate 100 µl of 10^{-1}, 10^{-2}, and 10^{-3} dilutions on SC+5-FOA and SC-arg+canavanine plates. To determine the number of viable cells, plate 100 µl of 10^{-4} and 10^{-5} dilutions on YPD plates. To determine the frequency of spontaneous mutation at the *CAN1* and *URA3* genes, prepare similar dilutions of the 4-2 haploid culture. For these cells, plate 100 µl of 10^{-1} and 10^{-2} dilutions on SC+5-FOA and SC-arg+canavanine plates. To determine the number of viable cells, plate 100 µl of 10^{-4} and 10^{-5} dilutions on YPD plates. Incubate all plates at 30°C.

(ii) *Meiotic recombination.* Transfer the 4-1 x 4-2 diploid from the YPD plate to liquid sporulation medium and incubate at 25°C according to Techniques and Protocols #9, Random Spore Analysis.

Day 9

(i) *Mutation and LOH.* Count the colonies on the selective plates and YPD plates and calculate the frequencies of Can^R and 5-FOA^R for the 4-1 x 4-2 diploid and for the 4-2 haploid. With the multipronged inoculating device (frogger), spot 22 independent isolates of each type of the drug-resistant strains onto SC+5-FOA, SC-arg+canavanine, SC-met (to score *HOM3*), and YPD plates. If you don't have enough 5-FOA^R or Can^R isolates of strain 4-2, just plate what you have. Include as controls the 4-1 x 4-2 diploid and the parental haploids 4-1 and 4-2.

(ii) *Meiotic recombination.* Check the sporulating culture of 4-1 x 4-2. If the culture has sporulated, digest and break apart tetrads, and plate dilutions of the random spores according to Techniques and Protocols #9, Random Spore Analysis.

Day 10

Mutation and LOH. On the basis of growth of patched or spotted strains, calculate the frequency of Can^R for strains first selected on 5-FOA and 5-FOA^R for strains first selected on canavanine.

Day 12

Using the multipronged device, patch or spot 45 independent random spore colonies from the 4-1 x 4-2 diploid onto YPD, SC-ura, SC-his, SC-met, and SC-arg+canavanine plates. Include as controls the 4-1 x 4-2 diploid and the parental haploids 4-1 and 4-2.

Day 15

Meiotic Recombination. Based on the growth of spotted spore colonies, determine the frequency of meiotic recombination between the *CAN1*, *URA3*, *HIS1*, and *HOM3* genes.

MATERIALS

Day 1   1 YPD plate

Day 3   1 YPD plate
1 SD+lys plate
Sterile loop or toothpicks
Sterile H$_2$O

Day 5   1 YPD plate

2 Culture tubes containing 5 ml of YPD

Day 6   Sterile microcentrifuge tubes

5 SC-arg+canavanine plates

5 SC+5-FOA plates

4 YPD plates

Liquid sporulation medium

Sterile H_2O

Glass spreader

70% Ethanol for sterilization

Day 9   2 SC-arg+canavanine plates

2 SC+5-FOA plates

2 SC-met

2 YPD plates

Sterile toothpicks/dowels

Multipronged inoculating device

2 Sterile 96-well plates

Materials for breaking apart tetrads from Techniques and Protocols #9, Random Spore Analysis

Day 12   1 SC-arg+canavanine plate

1 SC-ura plate

1 SC-met plate

1 SC-his plate

1 YPD plate

Sterile toothpicks/dowels

Multipronged inoculating device

1 Sterile 96-well plate

DATA   **Group #**

Spontaneous mutation (strain 4-2)

Medium	Dilution	# of Colonies	Freq. of Mutation
5-FOA	10^{-1}		
	10^{-2}		
CAN	10^{-1}		
	10^{-2}		
YPD	10^{-4}		
	10^{-5}		

Loss of heterozygosity (strain 4-1 x 4-2)

Medium	Dilution	# of Colonies	Freq. of LOH
5-FOA	10^{-1}		
	10^{-2}		
	10^{-3}		
CAN	10^{-1}		
	10^{-2}		
	10^{-3}		
YPD	10^{-4}		
	10^{-5}		

5-FOA^R (strain 4-1 x 4-2)

#	Can	Met
1		
2		
3		
4		
5		
6		
7		
8		
9		
10		
11		
12		
13		
14		
15		
16		
17		
18		
19		
20		
21		
22		
4-1		
4-2		
4-1 x 4-2		

Can^R (strain 4-1 x 4-2)

#	5-FOA	Met
1		
2		
3		
4		
5		
6		
7		
8		
9		
10		
11		
12		
13		
14		
15		
16		
17		
18		
19		
20		
21		
22		
4-1		
4-2		
4-1 x 4-2		

Meiotic recombination (random spore analysis)

#	Can^R	Ura	His	Met
1				
2				
3				
4				
5				
6				
7				
8				
9				
10				
11				
12				
13				
14				
15				
16				
17				
18				
19				
20				
21				
22				
4-1				
4-2				
4-1 x 4-2				

#	Can^R	Ura	His	Met
23				
24				
25				
26				
27				
28				
29				
30				
31				
32				
33				
34				
35				
36				
37				
38				
39				
40				
41				
42				
43				
44				
45				

Transformation of Yeast

Saccharomyces cerevisiae is unique among eukaryotes in the ease with which it can be transformed with DNA, and the high frequency with which the introduced DNA undergoes homologous recombination with genomic DNA. There are several requirements for a successful transformation experiment: (1) a means of introducing DNA into cells; (2) a selectable marker on the introduced DNA with corresponding nonreverting mutations in the chromosome; and (3) vector systems that allow propagation of cloned DNA in both *E. coli* and yeast.

TRANSFORMATION METHODS

The original method for yeast transformation involved incubating spheroplasted cells with DNA in the presence of polyethylene glycol (PEG) and $CaCl_2$ (Hinnen et al. 1978). A more convenient and much more widely used method involves the treatment of cells with the alkali salt, lithium acetate (LiAc), followed by incubation with DNA and PEG (Ito et al. 1983). DNA can also be introduced by electroporation (Hashimoto et al. 1985; Becker and Guarente 1991), whereby a brief electrical pulse permeabilizes the cells to DNA; by agitation of cells with glass beads (Costanza and Fox 1988); by bombardment of cells with DNA-coated particles (currently the only way to transform mitochondria) (Fox et al. 1988; Johnston et al. 1988); and by direct conjugation between bacterial and yeast cells (Heinemann and Sprague 1989). The method of choice depends on the purpose of the experiment, the number of strains to be transformed, and the desired number of transformants. The efficiency of transformation is often the most important parameter: If the goal is simply to put a plasmid into a given strain (only a few colonies needed), then any of the methods will work with almost any strain of yeast; if the goal is to get 10^6 transformants for screening a library, then the strain and transformation method must be carefully chosen. Spheroplasting, LiAc, and electroporation can all give high transformation frequency under optimal conditions.

SELECTABLE MARKERS

Although the frequency of transformation of yeast can be quite high, only a small fraction of the total number of cells in a transformation experiment become transformed.

Therefore, it is essential to have reliable selectable markers to select for those cells that have become transformed. The most common selectable markers used in yeast transformation are those that complement a specific auxotrophy. For example, the yeast *LEU2* gene encodes β-isopropylmalate dehydrogenase and complements the leucine auxotrophy of a *leu2* mutant. Other commonly used selectable markers are *URA3*, mutations in which result in uracil auxotrophy; *HIS3*, mutations in which result in histidine auxotrophy; and *TRP1*, mutations in which result in tryptophan auxotrophy. Of equal importance to the selectable marker is the corresponding chromosomal mutation that causes the auxotrophy. This mutation should be completely recessive and nonreverting; for example, the *leu2-3,112* mutation is a double frameshift mutation that reverts with a very low frequency ($<10^{-10}$) and is completely complemented by the wild-type *LEU2* gene. More recently, dominant drug resistance has been used as a selectable marker in yeast, as it is used in bacterial transformation (Hadfield et al. 1990). This has the potential advantage that the selectable marker lacks any homology with the yeast genome.

VECTOR SYSTEMS

Yeast cells that have taken up DNA during the transformation process can maintain that DNA, and thus become transformed, either by integration of the DNA into a chromosome or by autonomous replication. Integration into a chromosome takes place almost exclusively by homologous recombination in yeast. Once integrated, the transforming DNA is part of the chromosome and segregates in mitosis and meiosis with the same high fidelity as a chromosome. Plasmids used for integration have a yeast selectable marker, but no other yeast elements. Autonomous replication requires that the transforming DNA have a yeast origin of DNA replication. Originally called *ARS* (autonomously replicating sequence) elements, these can be either chromosomal DNA replication origins or the origin from the endogenous yeast 2 μ plasmid. Because the yeast replication origin is a relatively simple and short DNA sequence, DNA from other organisms will occasionally be found to have yeast *ARS* activity. To transform yeast cells stably, transforming DNA must either have sufficient homology to the yeast genome to integrate, or carry an *ARS* element.

Plasmids with only a chromosomal *ARS* element (ARS plasmid) have a variable copy number and often fail to segregate to the daughter cell in a division (Murray and Szostak 1983), resulting in a high rate of plasmid loss. Autonomously replicating plasmids may also have a centromere, or *CEN* element. A *CEN/ARS* plasmid (CEN plasmid) is more stable than a simple *ARS* plasmid because the centromere mediates the attachment of the plasmid to the mitotic spindle, ensuring segregation to both mother and daughter cells. Because of the high fidelity of segregation, the copy number is maintained at one or two plasmids per cell. CEN plasmids typically show a 2:2 (if the cell had one copy) or 4:0 (if the cell had at least two copies) segregation pattern in meiosis.

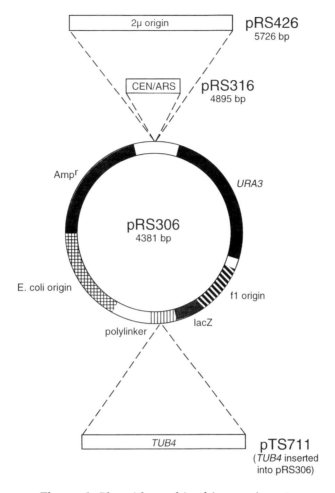

Figure 1. Plasmids used in this experiment.

An autonomously replicating plasmid that has the 2μ origin of replication (2μ-based plasmid) segregates in mitosis with about the same fidelity as a CEN plasmid, but is present at much higher copy number, typically 20–50 copies per cell. The high fidelity of segregation of 2μ plasmids depends on the presence of the endogenous 2μ plasmid. If a strain lacks the endogenous 2μ plasmid, then introduced 2μ-based plasmids will segregate as ARS plasmids (Murray and Szostak 1983).

Many older yeast plasmids were named systematically. In this system, integrating plasmids were designated YIp; ARS plasmids, YRp; CEN plasmids, YCp; and 2μ plasmids, YEp (E for *e*pisome). YRp plasmids are rarely used because of their extreme instability.

In addition to the yeast-specific elements, all standard yeast vectors also have a bacterial origin of replication and a bacterial selectable marker, usually ampicillin resistance. Older vectors (YIp5, YEp24, YCp50) are usually based on a pBR322 backbone. More recent plasmids are often based on plasmid backbones that replicate to higher copy number in bacteria, have useful polylinker sequences, and have single-stranded phage origins for the isolation of DNA for sequencing. A well-designed set of yeast vectors was described by Sikorski and Hieter (1989) and are used in this experiment. Figure 1 illustrates the plasmids to be used, all based on pRS306, an integrating vector with *URA3* as the selectable marker: pRS316 is pRS306 with an inserted *CEN/ARS* element, and pRS426 is a derivative of pRS 306 with an inserted 2μ origin (the orientation of the polylinker is also reversed). pTS711 is pRS306 with a genomic fragment including the yeast *TUB4* gene inserted in the polylinker.

EXPERIMENT V(a)

INTEGRATION

In most cases, integration of a circular plasmid into the yeast genome occurs by a single crossover and yields a direct repeat of the yeast sequence on the plasmid, as shown in Figure 2. Note that the entire plasmid is integrated, including the bacterial sequences and the yeast selectable marker—these now serve as a physical and phenotypic marker for the site of integration. The plasmid used in this experiment is pTS711, which has two regions of homology with the yeast genome: the *URA3* gene that is part of the parent vector, pRS306, and the *TUB4* gene that was inserted into the polylinker of pRS306. *TUB4* encodes β-tubulin, an essential protein involved in spindle pole body function (Marschall et al. 1996).

pTS711 cannot replicate in yeast because it lacks an *ARS* element and can, therefore, only transform yeast by integration. Integration can occur at either the *URA3* locus (actually the *ura3-52* locus in the strain used here) or the *TUB4* locus by homologous recombination (Fig. 2). Cutting a plasmid within a region of homology with the genome greatly increases the specificity and frequency of recombination at that site (Orr-Weaver et al. 1981); the free ends are highly recombinogenic and increase the efficiency of integration, and thus transformation, by approximately 100-fold. In this experiment, uncut pTS711 should transform strain 5-1 with a very low efficiency and integrate at either *ura3* or *TUB4*. Cutting pTS711 at the unique *Nco*I site in *URA3* should result in nearly 100% integration at *ura3*, and cutting pTS711 at the unique *Bgl*II site in *TUB4* should result in nearly 100% integration at *TUB4*.

Integration at a locus can be verified both genetically and physically. In this experiment, transformants will be crossed to tester strains to determine the site of integration. Integration at *TUB4* will be tested by crossing to a strain with a *tub4ts* mutation. If the plasmid integrates at the *TUB4* locus, then the plasmid-borne *URA3* gene will be

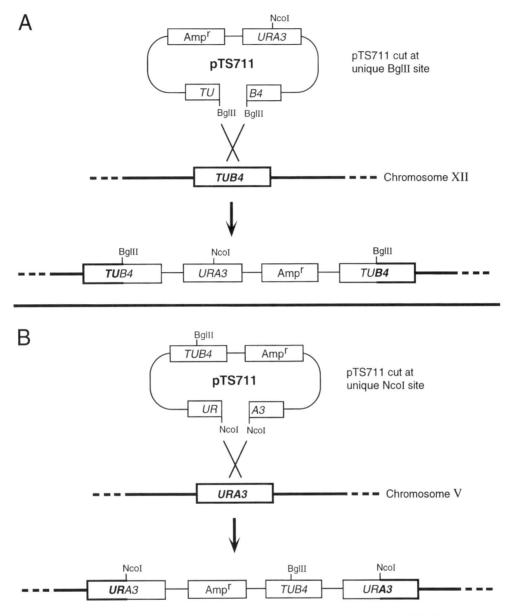

Figure 2. Integration of plasmid pTS711 at either the *TUB4* or *URA3* locus.

genetically linked to the *TUB4* locus; thus every tetrad should be a PD with two Ura⁺ Ts⁺ and two Ura⁻ Ts⁻ spores. Integration at *ura3* will be tested by crossing to a *URA3* strain. If the plasmid integrates at the *ura3* locus, then the plasmid-borne *URA3* gene will be genetically linked to the *ura3* locus; thus, every tetrad should be a PD with four Ura⁺ spores. As an exercise, predict the outcome of the cross if the plasmid had integrated at *TUB4* instead of *ura3*.

EXPERIMENT V(b)

REPLICATION AND STABILITY

Strain 5-1 will also be transformed with the autonomously replicating plasmids pRS316 and pRS426. The frequency of transformation will be higher for these plasmids than for the integrating plasmid. The stability of these plasmids in mitosis and meiosis will be examined.

STRAINS

5-1 TSY623   *MATα ade2-101 his3-Δ200 leu2-3,112 ura3-52*
5-2 TSY502   *MATa his3-Δ200 leu2-3,112 lys2-801 ura3-52 tub4-34*
5-3 TSY1017 *MATa his3-Δ200 leu2-3,112 trp1-1 ura3-52*
5-4 TSY808   *MATa lys2-801*

PLASMIDS

pTS711   *TUB4 URA3*
pRS316   *CEN URA3*
pRS426   *2μ URA3*

PROCEDURE

Day 1

Subclone strain 5-1 on a YPD plate. Incubate at 30°C.

Day 3

In the morning, inoculate 5 ml of YPD with a single colony of strain 5-1 to make a preculture. Incubate at 30°C with agitation. In the evening, determine the cell concentration using a hemocytometer. Assuming an approximate generation time of 100 minutes, calculate the volume of preculture to be added to 100 ml of YPD in order for the culture to reach a concentration of $2 - 10^7$ cells/ml by 10 a.m. on Day 4.

Day 4

Harvest the cells by centrifugation at 2000 rpm in a clinical centrifuge and follow the protocol for LiAc transformation given in Techniques and Protocols #1, High-efficiency Transformation of Yeast. Transform 5-1 under the conditions listed below.

Expt.	Recipient	DNA	Digestion	Selective medium
A	5-1	pTS711	none	SC-ura
	5-1	pTS711	NcoI	SC-ura
	5-1	pTS711	BglII	SC-ura
B	5-1	pRS316	none	SC-ura
	5-1	pRS426	none	SC-ura
Control	5-1	none	none	SC-ura

Day 7

(i) In order to analyze the transformants genetically, we will cross them to tester strains. Carefully pick two transformant colonies from each of the transformation plates with sterile toothpicks or a loop and mix them with the tester strains in a small patch on a YPD plate. Cross the strains in the combinations shown below, and incubate overnight at 30°C (for crosses involving 5-2, which contains a ts mutation, incubate at room temperature).

Transformant	Tester	Selective medium
Integration location		
5-1+pTS711(NcoI)	5-4	SD
5-1+pTS711(NcoI)	5-2	SD+his+leu
5-1+pTS711(BglII)	5-4	SD
5-1+pTS711(BglII)	5-2	SD+his+leu
Meiotic stability		
5-1+pTS711(BglII)	5-3	SD+his+leu
5-1+pRS316	5-3	SD+his+leu
5-1+pRS426	5-3	SD+his+leu

(ii) The assays for mitotic stability require that transformants be purified away from the untransformed cells on the transformation plates. Purify two transformants each from the BglII-cut integrating transformation (Expt. V[a] on Day 4) and the autonomously replicating plasmid transformations (Expt. V[b] on Day 4) by streaking onto SC-ura agar. Incubate at 30°C.

Day 8

Replica-plate the YPD plates containing the crosses onto the appropriate selective medium. Only diploid cells formed by mating should be able to grow in the patches where the cells were mixed. Incubate at 30°C.

Day 9

(i) In preparation for sporulation, patch the diploids growing on the selective plates to a YPD plate and incubate for 1 day at room temperature.

(ii) Inoculate one 5-ml YPD culture with a toothpickful of cells from one colony of each purified transformant from the SC-ura plates. Place the tubes on the roller drum overnight at 30°C.

Day 10

(i) Transfer the diploids from YPD to sporulation medium (both liquid and plate to compare efficiency of sporulation) and incubate several days at room temperature.

(ii) Make serial dilutions of the overnight 5-ml YPD cultures in sterile H_2O and spread aliquots on YPD plates. Try to get 100–300 colonies/plate. Incubate at 30°C.

Day 12

(i) Begin dissecting tetrads from the crosses involving the transformants. Try to dissect ten tetrads from each cross. Incubate tetrad plates at 30°C, <u>except</u> for those involving 5-2, which should be incubated at room temperature.

(ii) Replica-plate YPD plates with the appropriate number of colonies to SC plates and SC-ura plates. Incubate at 30°C.

Day 13–14

(i) As tetrads grow up on the dissection plates, replica-plate the crosses to strains 5-3 and 5-4 onto SC-ura and YPD plates at 30°C to determine segregation of *URA3*. For the crosses of the pRS316 transformants to strain 5-3, replica-plate to SC-trp to determine centromere segregation. Replica-plate crosses to 5-2 onto YPD plates at both room temperature and 37°C to determine segregation of *TUB4*.

(ii) Score the fraction of cells that remained Ura⁺ after nonselective growth in YPD.

MATERIALS

Day 1   1 YPD plate

Day 3   1 Culture tube containing 5 ml of YPD
                  Erlenmeyer flask containing 100 ml of YPD

Day 4 Materials for Techniques and Protocols #1, High-efficiency
Transformation of Yeast
7 SC-ura plates

Day 7 3 YPD plates
3 SC-ura plate

Day 8 1 SD plate
2 SD+his+leu plates

Day 9 2 YPD plates
3 Culture tubes, each containing 5 ml of YPD

Day 10 15 Culture tubes, each containing 5 ml of sterile H_2O
12 YPD plates
7 Culture tubes, each containing 2 ml of sporulation medium
2 sporulation plates

Day 12 10 YPD plates for tetrad dissection
6 SC plates
6 SC-ura plates

Day 13 7 SC-ura plates
7 YPD plates
2 SC-trp plates

REFERENCES

Becker D.M. and Guarente L. 1991. High-efficiency transformation of yeast by electroporation. *Methods Enzmol.* **194:** 182–187.

Costanza M.C. and Fox T.D. 1988. Transformation of yeast by agitation with glass beads. *Genetics* **120:** 667–670.

Fox T.D., Sanford J.C., and McMullin T.W. 1988. Plasmids can stably transform yeast mitochondria lacking endogenous mtDNA. *Proc. Nat. Acad. Sci.* **85:** 7288–7292.

Hadfield C., Jordan B.E., Mount R.C., Pretorius G.H.J., and Burak E. 1990. G418-resistance as a dominant marker and reporter for gene expression in *Saccharomyces cerevisiae. Curr. Genet.* **18:** 303–314.

Hashimoto H., Morikawa H., Yamada Y., and Kimura A. 1985. A novel method for transformation of intact yeast cells by electroinjection of plasmid DNA. *Appl. Microbiol. Biotechnol.* **21:** 336–339.

Heinemann J.A. and Sprague G.F. 1989. Bacterial conjugative plasmids mobilize DNA transfer between bacteria and yeast. *Nature* **340:** 205–209.

Hinnen A., Hicks J.B., and Fink G.R. 1978. Transformation in yeast. *Proc. Natl. Acad. Sci.* **75:** 1929–1933.

Ito H., Fukuda Y., Murata K., and Kimura A. 1983. Transformation of intact yeast cells treated with alkali cations. *J. Bacteriol.* **153:** 163–168.

Johnston S.A., Anziano P., Shark K., Sanford J.C., and Butow R.A. 1988. Transformation of yeast mitochondria by bombardment of cells with microprojectiles. *Science* **240:** 1538–1541.

Marschall L., Jeng R., Mulholland J., and Stearns T. 1996. Analysis of Tub4p, a yeast γ-tubulin-like protein: Implications for microtubule organizing center function. *J. Cell Biol.* **134:** 443–454.

Murray A.W. and Szostak J.W. 1983. Pedigree analysis of plasmid segregation in yeast. *Cell* **34:** 961–970.

Orr-Weaver T., Szostak J., and Rothstein R. 1981. Yeast transformation: A model system for the study of recombination. *Proc. Nat. Acad. Sci.* **78:** 6354–6358.

Sikorski R.S. and Hieter P. 1989. A system of shuttle vectors and yeast host strains designed for efficient manipulation of DNA in *Saccharomyces cerevisiae. Genetics* **122:** 19–27.

Synthetic Lethal Mutants

Synthetic lethal mutants are often used to identify genetic interactions (Huffaker et al. 1987; Guarente 1993; Appling 1999). The goal is to identify a double mutant combination that is inviable when both of the single mutants alone are viable. A synthetic lethal screen can begin with a deletion mutation in a nonessential gene or with a mutation in an essential gene that does not eliminate function. A common example, in the latter case, is to begin with a temperature-sensitive mutant and perform the synthetic lethal screen at the permissive temperature.

We will begin with a strain that is deleted for the nonessential *BUB2* gene, one of the genes originally identified as a component of the spindle checkpoint pathway. Bub2p is part of a two-component GTPase activating protein (GAP) that regulates the activity of a small G-protein called Tem1p. Tem1p is involved in regulating the exit from mitosis. Cells treated with the drug benomyl arrest in mitosis under the control of an intracellular signal transduction pathway called the spindle checkpoint. *BUB2* is required to keep cells arrested in the cell cycle in response to benomyl. We will perform a synthetic lethal screen using a *bub2* deletion mutant as a starting strain. We induce mutations and look for double mutants that are inviable. The mutants will identify genes that are essential in the absence of Bub2p.

We will identify the inviable double mutant using a counter-selectable YCp plasmid PRS318 that contains *LEU2* and *CYH2*. Resistance to the translation inhibitor cycloheximide is conferred by recessive mutations in the *CYH2* gene that encode ribosomal protein L28. Therefore, a *leu2 cyh2* strain transformed with pRS318 is a leucine prototroph and sensitive to cycloheximide. Plating the strain onto medium containing cycloheximide selects for cells that lose the plasmid. The rate of plasmid loss for pRS318 is about 1% per generation. Therefore, colonies from a wild-type strain produce many cells that lose the plasmid during growth on a plate. If the colony is replica-plated to medium containing cycloheximide, you select for cells within the colony that have lost the plasmid. Such colonies have a mottled appearance, and the strain is said to "papillate" to cycloheximide resistance. We will begin with a *bub2 leu2 cyh2* strain transformed with pRS318 containing *BUB2*. Colonies can papillate to cycloheximide resistance because *BUB2* is nonessential. We will mutagenize and screen for colonies that cannot produce cycloheximide-resistant cells and papillate to cycloheximide resistance. These are double mutants that are inviable in the absence of *BUB2*. To facilitate the analysis, we will mutagenize both mating types so that we can quickly order mutants into complementation groups.

The basic principle for the synthetic lethal screen is shown in Figure 1. The strain on the left is the starting strain that has the *BUB2* gene deleted by replacing the coding sequence with *URA3* (*bub2::URA3*). The plasmid contains *BUB2*, *LEU2*, and *CYH2*. The strain on the right is identical except for the mutation (*mut*) we induced in a gene that is unlinked to *BUB2*. Shown below the strains are colonies grown on YPD plates. The white color represents good growth of cells within the colony on YPD plates. Note that cells losing the plasmid can grow on YPD plates. Replica-plating the starting strain to YPD plus cycloheximide results in patchy growth of the colony (papillation). Each papilla is the result of a plasmid loss event during growth on YPD plates. The strain on the right cannot generate papillae when the colony is replica-plated from a YPD plate to a YPD plate containing cycloheximide. Therefore *mut* and *bub2::URA3* are synthetically lethal. The *bub2::URA3* strain is viable and can lose the plasmid. The *mut* strain is only viable when the *bub2::URA3* is complemented by the plasmid.

STRAINS

6-1 2405   *MATa ade5 his3 leu2 ura3 bub2::URA3 cyh2 [pMCM90]*
6-2 2406   *MATα lys2 his3 leu2 ura3 bub2::URA3 cyh2 [pMCM90]*

PLASMID

pMCM90 [YCp *LEU2 BUB2 CYH2*]

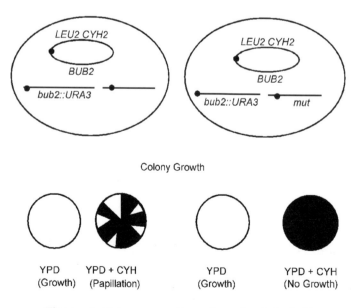

Figure 1. Colony growth and synthetic lethality.

PROCEDURE

SAFETY NOTE

EMS is a potent mutagen. Wear gloves and work in a hood when tubes are open. Neutralize EMS with 5% sodium thiosulfate before discarding. Use disposable pipettes for all manipulations.

Day 1

You will be given a 5-ml culture of strain 6-1 or strain 6-2, previously grown to saturation in HC-leu. Transfer 1 ml of cells to a sterile microfuge tube and wash twice with sterile H_2O. Resuspend in 1 ml of sterile phosphate buffer, pH 7.0. Add 30 μl of EMS to the cells and disperse by agitation. Incubate for 1 hour at 30°C with agitation. Centrifuge and resuspend in sterile H_2O. Transfer to a new tube and wash twice with 5% sodium thiosulfate. Resuspend cells in 1 ml of HC-leu. Add the mutagenized cells to 4 ml of HC-leu and incubate 2 days at 30°C.

Day 3

Count cells in a counting chamber and dilute cells to 10^3/ml in sterile H_2O. Plate 0.2 ml of cells per plate onto 20 HC-leu plates and grow for 3 days at 30°C.

Day 6

Replica-plate to YPD plates containing cycloheximide and incubate 2 days at 30°C. Keep the master plates (HC-leu) and place them back at 30°C.

Day 8

Identify colonies that cannot produce cycloheximide-resistant papillae. Reclone candidates from the YPD master plates onto new YPD plates at 30°C.

Day 10

Exchange mutants with all others to get a collection in both mating types. Label the strains and record the source (the group that identified the mutant). Streak all mutants of similar mating type in a straight line on YPD plates to do crosses. You should have at least two plates, one with mutants derived in strain 6-1 and one with mutants derived in strain 6-2. Be sure to include an unmutagenized parent, 6-1, and 6-2 on the appropriate plates. We will determine dominance for the synthetic lethality by selecting diploids using the mutants and the unmutagenized parents. Leave overnight at 30°C.

Day 11

Replica-plate the strains to cross them on YPD. Incubate overnight at 30°C.

Day 12

Replica-plate to SD + his + leu plates to select for diploids. Incubate overnight at 30°C.

Day 13

Replica-plate to SD + his + leu containing cycloheximide. Incubate overnight at 30°C.

Day 14

Score the dominance and determine the number of complementation groups.

MATERIALS

Day 1	1 Sterile microfuge tube
	Sterile phosphate buffer, pH 7.0
	EMS (methanesulfonic acid, ethyl ester; Sigma M0880)
	Sterile 5% sodium thiosulfate (w/v)
	Sterile YPD
	1 Sterile culture tube
Day 3	Sterile dH$_2$O
	20 HC-leu plates
Day 6	20 YPD plates containing cycloheximide
Day 8	~5 YPD plates
Day 10	~4 YPD plates
Day 11	2 YPD plates
Day 12	2 SD + his + leu plates
Day 13	2 SD + his + leu + cycloheximide plates

REFERENCES

Appling D.R. 1999. Genetic approaches to the study of protein-protein interactions. *Methods* **19:** 338–349.

Guarente L. 1993. Synthetic enhancement in gene interaction: A genetic tool comes of age. *Trends Genet.* **9:** 362–366.

Huffaker T.C., Hoyt M.A., and Botstein D. 1987. Genetic analysis of the yeast cytoskeleton. *Annu. Rev Genet.* **21:** 259–284.

Gene Replacement

One of the most powerful and important techniques available for studies in yeast is gene replacement. This technique allows the replacement of a gene at its normal chromosomal location with an allele of that gene created in vitro, such that the only genetic difference between the initial strain and the final strain is that particular allele. Using this method, phenotypes conferred by null mutations or any other types of mutations made in a cloned gene can be analyzed. In theory and in practice, a cloned yeast gene can be changed and then recombined into the genome, precisely replacing the wild-type allele.

Determination of the null phenotype for a gene is an essential step in understanding the function of that gene. First, it reveals whether the gene is essential for growth. Second, if the gene is not essential for growth, it allows study of strains completely lacking that particular function. To examine the phenotype conferred by a null mutation, gene replacement is generally done in diploid strains, in case the null mutant is inviable as a haploid. The inviability will be observed after tetrad analysis of a sporulated culture. If the gene is essential, viability will segregate 2:2 in the tetrads. If the gene is not essential, all four spores will be viable and two will be null mutants for that gene.

EXPERIMENT VII(a)

ONE-STEP GENE DISRUPTION

Traditionally, a one-step gene replacement (Rothstein 1983) has been done by transformation with a restriction fragment that contains the mutant allele and has ends homologous to the locus where integration is desired. However, in order to create the mutant allele fragment, the gene of interest had to be cloned, and then using convenient restriction sites, the gene, or a portion of it, was replaced by a DNA fragment encoding a selectable marker in yeast. Because of the inherent limitations of these manipulations, it is frequently difficult to create a true null allele, in which the entire coding sequence of a gene is precisely removed. However, it is now much easier to make null mutations in yeast using the PCR-mediated gene disruption method (Baudin et al. 1993; Lorenz et al. 1995).

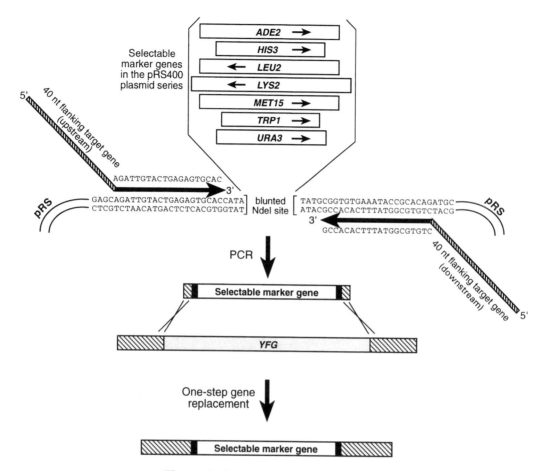

Figure 1. One-step gene replacement.

PCR-mediated gene disruption is based on the fact that homologous recombination in yeast is very efficient with linear DNA fragments, and that only ~40 bp of homology is required for rather efficient recombination. The availability of the entire *S. cerevisiae* genome sequence in combination with PCR-mediated gene disruption has made it possible to create a null mutation of any gene without ever having to clone it. Furthermore, a series of plasmids and strains have been created that increase the efficacy of this technique (Brachmann et al. 1998). The plasmids contain common yeast selectable marker genes cloned into a conserved site. This permits one set of PCR primers to be used to disrupt a gene with a choice of markers (see Fig. 1). The yeast strains have the common auxotrophic markers completely deleted from the genome, which eliminates the unwanted background of gene conversion events that can happen between the selectable markers and their mutant alleles in the chromosome.

Often, gene disruptions are carried out to test the phenotype of a null allele of a gene of interest. In this experiment, we will disrupt the gene for a protease in order to make a strain that will be useful for biochemical experiments. One of the difficulties in doing biochemistry on yeast cell extracts is the abundance of proteases contained within the vacuole that are released when cells are broken open. A truly protease-deficient strain requires disruption of each of the different genes that encode vacuolar proteases. Fortunately, there is a way to eliminate most of the proteases in a single step. Vacuolar proteases are synthesized as inactive proenzymes that become active in the vacuole on proteolytic removal of the propeptide. The PEP4 gene encodes the major processing protease, proteinase A. Disruption of *PEP4* prevents all of the other proteases from becoming active. A complication of this procedure is that proteinase B can also cleave propeptides and activate vacuolar proteases. In the absence of proteinase A, active proteinase B can continue to process vacuolar proteases for several cell divisions, but the self-activation of proteinase B is not sustained indefinitely. Once the activity of proteinase B falls below a threshold, the cell becomes irreversibly protease-negative. After we disrupt *PEP4*, we will streak out the disruption strain for single colonies several times to dilute proteinase B and to obtain a protease-negative strain.

A different PCR method will be used to screen transformed colonies for a disrupted *PEP4* gene. In this procedure, screening will be done on colonies directly, without the need to purify DNA from yeast. Transformants that are genotypically *pep4* will then be screened, using a plate assay, for Carboxypeptidase Y activity. This peptidase is a vacuolar enzyme that requires proteinases A and B for activation.

STRAIN

7-1   BY4732   *MATα his3Δ200 met15Δ0 trp1Δ63 ura3Δ0*

PLASMID

pRS303   An *E. coli* vector carrying the yeast *HIS3* gene that can be used as a template for PCR-mediated gene disruption.

PROCEDURE

Day 1

Streak strain 7-1 on a YPD plate. Incubate at 30°C.

Day 3

Pick a robust colony of strain 7-1 and use it to inoculate a 5-ml YPD culture. Incubate overnight at 30°C.

Day 4

Determine the cell density of the 5-ml strain 7-1 culture using the hemocytometer (see Appendix G). Dilute cells to 5×10^6 cells/ml in an Erlenmeyer flask containing 50 ml of YPD and grow culture for two additional divisions at 30°C (about 4 hours).

Harvest the cells and transform them, following Techniques and Protocols #1, High-efficiency Transformation of Yeast, with ~1 μg of the *pep4::HIS3* PCR-generated DNA (a stock will be provided that was made via Techniques and Protocols #13, PCR Protocol for PCR-mediated Gene Disruption, using RS303 as a vector template, and the two primers:

5′PEP4DEL   TGGTCAGCGCCAACCAAGTTGCTGCAAAAGTCCACAAGG*CAGATTG-TACTGAGAGTGCAC*

3′PEP4DEL   AATCGTAAATAGAATAGTATTTACGCAAGAAGGCATCACC*CTGTGCG-GTATTTCACACCG*). Also be sure to carry out a transformation with <u>no</u> DNA, as a negative control. Plate each transformation onto one HC-his plate. Incubate for 3 days at 30°C.

Day 7

Pick 11 transformants and streak for single colonies on HC-his plates. Use only three plates and streak 3–4 transformants per plate. Incubate for 2 days at 30°C.

Day 9

(i) Pick one colony from each of the 11 transformants that were streaked on HC-his (from Day 7) and determine whether the *PEP4* gene has been disrupted using Techniques and Protocols #14, Yeast Colony PCR Protocol. Also make certain to include a sample of strain 7-1 cells as a control. Amplify the DNA by using the two oligos:

      5′PEP4   GCTTGAAAGCATTATTGCCATTGGCC

      3′PEP4   GGCCAAACCAACCGCATTGTTGCCC

(ii) While the PCR amplification is being carried out, prepare a 1.5% agarose gel.
Check the PCR products from each colony by gel electrophoresis. Add 3 μl of agarose gel loading buffer to each PCR sample and load all of each sample on the gel. Make certain to include a DNA size marker. The wild-type *PEP4* gene should produce an ~1.2 kb band, whereas the *pep4::HIS3* allele has a predicted size of ~1.4 kb.

(iii) Streak out three of the *pep4::HIS3* transformants as well as strain 7-1 onto YEPD plates. Streak two strains per plate. Incubate for 3 days at 30°C.

Day 12

Assay the colonies on the YEPD plates using the APE protease plate assay as in Techniques and Protocols #8, Plate Assay for Carboxypeptidase Y. Note any differences in color development between colonies.

MATERIALS

Day 1	1 YPD plate
Day 3	1 Culture tube containing 5 ml of YPD
Day 4	Materials for Techniques and Protocols #1, High-efficiency Transformation of Yeast Erlenmeyer flask containing 50 ml of YPD 2 HC-his plates
Day 7	3 HC-his plates

Day 9	Materials for Techniques and Protocols #14, Yeast Colony PCR Protocol
	~120 pmoles of each DNA oligo:
	   5´PEP4 GCTTGAAAGCATTATTGCCATTGGCC
	   3´PEP4 GGCCAAACCAACCGCATTGTTGCCC
	Materials and equipment for a 1.5% agarose gel
	2 YPD plates

Day 12	Materials for Techniques and Protocols #8, Plate Assay for Carboxypeptidase Y

EXPERIMENT VII(b)

TWO-STEP GENE REPLACEMENT

The second type of gene replacement commonly used is a two-step method (Scherer and Davis 1979) in which one first integrates a plasmid that contains the mutant allele. When the integration occurs at the locus of interest, it will result in a duplication of the region—one duplicate will be wild-type and the other mutant—with the plasmid sequences in between. Homologous crossovers between the copies from the duplication will result in excision of the plasmid and loss of one of the two copies of the dupli-

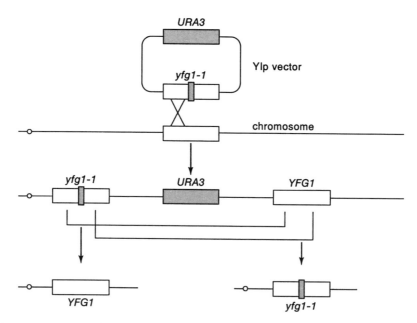

Figure 2. Two-step gene replacement. Integration of a plasmid by a single crossover results in a duplication of the *YFG1* locus. One copy contains the wild-type *YFG1* gene and the other copy contains the *yfg1-1* mutation. Strains that have excised the plasmid can be selected using 5-FOA medium. Depending on the location of the crossover, the excision will leave behind either *YFG1* or *yfg1-1*.

cated region. Depending on the exact location of the crossover, the copy left behind will contain either the mutant or wild-type form of the gene (Fig. 2). These can be distinguished by phenotype and/or by Southern hybridization analysis. Excision of the plasmid is done most conveniently if the selectable marker on the plasmid is URA3; one can simply select for those that have lost the plasmid using 5-fluoro-orotic acid (5-FOA) plates.

Two-step gene replacement is the method of choice under two conditions:

1. If the desired mutation is not associated with a selectable phenotype, direct selection by the one-step method is not possible. This is the case for gene replacement with a temperature-sensitive allele of a particular gene. Using the two-step method, one can screen recombinants that have looped out the plasmid for those containing the mutant allele.

2. Using the two-step method, one can generate both the wild-type and mutant strains from the same initial transformant strain. This provides strains that are isogenic except in the gene of interest. A two-step gene replacement protocol will be used in Experiment IX to switch mating-type alleles at the MAT locus.

EXPERIMENT VII (c)

THE PLASMID SHUFFLE

In addition to analyzing phenotypes through gene replacement, it is possible to screen for mutant phenotypes when the gene of interest is on an autonomous plasmid, particularly when attempting to identify conditional lethal alleles of an essential gene, such as your favorite gene (*YFG*). Using the plasmid-shuffle technique (Sikorski and Boeke 1991), a plasmid containing *YFG* is mutagenized, the pool of mutagenized plasmids is transformed into yeast and then screened for a conditional mutant phenotype in a strain that contains a *yfg* deletion mutation. This is done by transforming a strain that already contains: (1) the wild-type gene of interest on an autonomous plasmid with a counter-selectable gene (such as *URA3*) as the selectable marker (the counter-selectable gene has both positive and negative selection and you can select cells that have the plasmid and cells that do not have the plasmid) and (2) a *yfg* null allele in the genome.

In such a strain, viability is dependent on the *YFG* encoded on the plasmid. A second plasmid with a different selectable marker is mutagenized (usually in vitro) and transformed into your strain. Replica plating, to select against the plasmid containing *YFG*, identifies potential mutants. For *URA3*-containing plasmids, the selection is done using 5-FOA plates under different conditions (e.g., temperature, medium conditions). Only cells that have lost the wild-type gene on the *URA3* plasmid can grow on the 5-FOA plates, and the phenotype conferred by the mutant plasmid will be expressed.

In this experiment, we will look for temperature-sensitive and benomyl-sensitive alleles of the essential *NDC10* gene that encodes a component of the kinetochore (*NDC10=YFG*). Two libraries of mutagenized plasmids will be provided, one mutagenized by hydroxylamine (Techniques and Protocols #6) and the other by passage through *E. coli* strain XL-1 Red (*Stratagene*). The XL-1 red strain is a *mutS, mutD, mutT* triple mutant. The three mutations almost completely eliminate DNA repair: *mutS* lacks error-prone mismatch repair, *mutD* lacks the 3′-5′ exonuclease of DNA polymerase III and *mutT* lacks 8-oxodGTP hydrolysis. Plasmids are transformed into the XL-1 Red cells and then the strain is propagated overnight to accumulate mutations. The plasmids are recovered and amplified to give a library of mutagenized plasmids. The mutagenized plasmid pDB141 is a YCp *LEU2 NDC10* plasmid. Strain 7-2 has an *ndc10*::TRP1 allele that is viable because it contains pRG68, a YCp *URA3 NDC10* plasmid that will be used for counterselection with 5-FOA as described above.

STRAIN

7-2 2404   *MATα ade2 trp1 leu2 ura3 his3 lys2-801 ndc10::TRP1* [pRG68]

PLASMIDS

pDB141	A YCp *LEU2* plasmid containing *NDC10*
pRG68	A YCp *URA3* plasmid containing *NDC10*

PROCEDURE

Day 1

Streak strain 7-2 onto a YPD plate. Incubate at 30°C.

Day 3

Pick a robust colony of strain 7-2 and use it to inoculate a 5-ml YPD culture. Incubate overnight at 30°C.

Day 4

Determine the cell density of the 5-ml strain 7-2 culture using the hemocytometer (see Appendix G). Dilute cells to 5×10^6 cells/ml in 50 ml of YPD and grow the culture for two additional divisions at 30°C (about 4 hours). Harvest the cells and transform them, following Techniques and Protocols #1, High-efficiency Transformation of Yeast, with 100 ng of the mutagenized pDB141 plasmid (a stock will be provided that was made either via Techniques and Protocols #6, Hydroxylamine Mutagenesis of Plasmid DNA or by passage in XL1-Red cells). Make certain to include a no-DNA control. There will be a small modification to Techniques and Protocols #1, High-efficiency Transformation of Yeast. Modify step 16 by resuspending the transformed cells in 2.5 ml of sterile distilled H₂O (concentrated cells). Remove 0.4 ml of the concentrated cells and dilute this to 2.0 ml (diluted cells). Plate 0.2-ml aliquots onto HC-leu plates, using ten plates for the concentrated cells and another ten plates for the diluted cells. For the no-DNA control, plate a 0.2-ml aliquot of concentrated cells on one plate. Incubate all 21 plates at 23°C.

Day 9

Select ten transformation plates that ideally have 100–300 colonies each and determine the total number of colonies that will be assayed. Replica-plate each of the ten HC-leu plates onto 5-FOA plates. Keep the master plates (HC-leu). Incubate at 23°C.

Day 11

Determine the fraction of colonies that cannot papillate to FOA resistance. Replica-plate the FOA plates to two YPD plates and one YPD plate containing 15 μg/ml of benomyl (BEN). Incubate one YPD plate at plate at 37°C, and the YPD plate and the YPD + BEN plate at 23°C.

Day 12

Score the replica plates at 37°C for temperature-sensitive candidates. Pick as many as 12 temperature-sensitive colonies from the YPD plate incubated at 23°C and purify by streaking onto YPD at 23°C (use a maximum of three YPD plates).

Day 13

Score the BEN plates at 23°C for colonies that are benomyl sensitive and unable to grow or grow poorly. Pick as many as 12 candidates and purify by streaking on YPD at 23°C (use a maximum of three YPD plates). Note any benomyl-sensitive colonies that are also temperature-sensitive.

Day 16

Retest the purified candidates for benomyl sensitivity on YPD + BEN and temperature sensitivity on two YPD plates, incubating one at 23°C and the other at 37°C.

Day 17

Score the plates at 37°C for temperature sensitivity. Score the benomyl plates at 23°C and keep the benomyl plates until tomorrow.

Day 18

Score the benomyl plates and determine the frequency of each class of mutants.

MATERIALS

Day 1   1 YPD Plate

Day 3   Culture tube containing 5 ml of YPD

Day 4   Materials for Techniques and Protocols #1, High-efficiency
                 Transformation of Yeast Mutagenized pDB141 DNA
             50 ml of YPD in a flask
             21 HC-leu plates

Day 9   10 5-FOA plates
              Sterile velveteen pads

Day 11   20 YPD plates
              10 YPD plates containing 15 μg/ml benomyl
              Sterile velveteen pads

Day 12   3 YPD plates

Day 13   3 YPD plates

Day 16   3 YPD + BEN plates
              6 YPD plates
              Sterile velveteen pads

Experiment VII (d)

CONSTRUCTING PROTEIN FUSIONS

Epitope tagging is a powerful technique that is used to analyze proteins in yeast. A protein fusion is made using a peptide sequence for which there are often commercially available antibodies. There are a number of different epitopes that can be used. There are small peptide sequences that are highly antigenic such as the HA peptide from the hemagglutinin protein of influenza virus, a peptide (MYC) from the c-Myc protein and the FLAG epitope. Fusions with small proteins also function as epitope tags and can have even greater utility. For example, GFP (green fluorescent protein) is used to localize fusion proteins in vivo and GST (glutathione-*S*-transferase) is used to affinity-purify fusion proteins.

A number of different methods can be used to engineer protein fusions. The small peptides like HA or MYC can be attached to either end of the protein or can be inserted in-frame in the coding sequence. Small protein fusions, like GFP or GST, are usually made at either terminus of the protein. It is important to show, if at all possible, that the protein fusion to your favorite gene functions normally. This is best accomplished by complementing all mutant phenotypes associated with deletions of your favorite gene.

GFP FUSION

We will illustrate protein fusions using two different methods. The first will use a PCR-mediated one-step gene replacement (Longtine et al. 1998). We will generate a carboxy-

Primer Design for Ndc10-GFP fusion

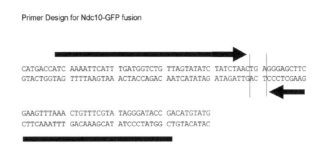

```
CATGACCATC AAAATTCATT TGATGGTCTG TTAGTATATC TATCTAACTG AGGGAGCTTC
GTACTGGTAG TTTTAAGTAA ACTACCAGAC AATCATATAG ATAGATTGAC TCCCTCGAAG


GAAGTTTAAA CTGTTTCGTA TAGGGATACC GACATGTATG
CTTCAAATTT GACAAAGCAT ATCCCTATGG CTGTACATAC
```

Figure 3. Primers for carboxy-terminal fusion to Ndc10p.

terminal GFP fusion of an essential gene encoding a kinetochore protein. Kinetochores are localized near spindle poles through most of the cell cycle and therefore have a staining pattern similar to the Tub4-GFP fusion that we have already seen. There is one important difference in the pattern of staining. Kinetochores move from the poles onto the spindle at mitosis and stain the intranuclear microtubules of the short (1–2 μm) pre-anaphase spindle.

The rationale is identical to the PCR strategy used in Experiment VII (a), except that the selectable marker is the *HIS3* homolog from the yeast *Schizosaccharomyces pombe*. The *S. pombe his5⁺* gene will complement *his3* mutants of *S. cerevisiae* and is used to reduce the possibility of gene conversion adding to the background of His⁺ colonies after transformation. The primer design for the experiment is shown in Figure 3. Forty base pairs 5′ of the stop codon (arrow) are used to make the 5′ end of the NDCTAG1 primer. In addition, the following sequence from the 5′ end of the GFP gene in plasmid pFA6a-GFP(S65T)-*HIS3*MX6 is added to the 3′ end of the NDCTAG1 primer:

GCGGGTGCTGGAGCA**ATG**AG

The start codon (ATG) for GFP is in bold. Forty base pairs 3′ of the stop codon are used to make the 5′ end of NDCTAG2 primer. In addition, the following sequence from the 3′ end of the *S. pombe his5⁺* gene in plasmid pFA6a-GFP(S65T)-*HIS3*MX6 is added to the 3′ end of the NDCTAG2 primer:

CCTCGAGGTCGACGGTATC

The sequences at the 3′ end of each primer are used to prime the PCR amplification from plasmid pFA6a-GFP(S65T)-*HIS3*MX6. The scheme is shown diagrammatically in Figure 4. The two 5′ ends of the primers direct the integration to make the precise GFP fusion at the carboxyl terminus of the Ndc10 protein.

To confirm the integration event, we will use a PCR primer from the 3′ end of the *NDC10* gene in combination with a primer from the 5′ end of GFP that will produce a 250-bp product (Primers 1 and 2 in Fig. 4). We will also use a primer from the 3′ end of *his5⁺* in combination with a primer from the 3′ flanking sequences of *NDC10* that will produce a 350-bp product (Primers 3 and 4 in Fig. 4). The PCR assays are done

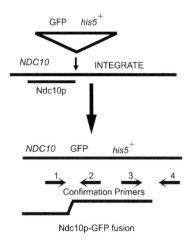

Figure 4. Ndc10p–GFP fusion.

together. A productive integration produces two PCR products of 250 bp and 350 bp. In the absence of integration, the *NDC10* gene remains intact and PCR produces a single product of 600 bp.

DEGRON FUSION

The second method will use integrative transformation with a YIp plasmid to generate an amino-terminal "degron" fusion to a different kinetochore protein, Cep3p. Degron tagging is used to construct temperature-sensitive mutations in essential genes (Dohmen et al. 1994). A degron is a complex amino-terminal fusion that targets the associated protein for proteolysis by the 26S proteasome. The fusion has the design shown in Figure 5.

The *CUP1* promoter regulates the fusion gene so that expression is dependent on the presence of copper in the medium. The degron has a single ubiquitin protein at the extreme amino terminus that provides the initiation codon for translation (M). The ubiquitin is rapidly cleaved upon expression, and the cleavage occurs immediately after the ubiquitin peptide to uncover an arginine (R) residue at the amino terminus of the resulting fusion protein. The arginine is a destabilizing residue for the N-end rule and targets the protein for degradation when internal lysine residues are exposed. A mutant

Figure 5. Degron-tagged genes.

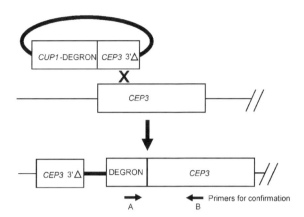

Figure 6. Degron-tagged *CEP3*.

dihydrofolate reductase protein from mouse (mDHFR^{ts}) unfolds at 37°C to expose lysines and target the entire fusion protein for proteolysis within the 26S proteasome. The 26S proteasome is highly efficient and, therefore, the mutant phenotype results from a complete lack of protein. This is easily confirmed because the fusion contains an internal HA epitope tag. The fate of the protein is determined by Western blots. The result is a fusion that generates a temperature-sensitive mutant, which grows well at low temperatures (23–25°C) and is inviable at 37°C. Some degron-tagged mutants are somewhat "leaky" and grow slightly under restrictive conditions. This can usually be overcome by forcing the expression of excess Ubr1p, a ubiquiting protein ligase.

Degron-tagged temperature-sensitive alleles may be preferable to temperature-sensitive alleles constructed by in vitro mutagenesis. Degron tagging produces loss-of-function or null alleles. In vitro mutagenesis produces predominantly missense mutations, and often there is residual protein activity, even at the restrictive temperature. The residual activity can sometimes confound the phenotype. There are two other utilities to the degron construct. The HA epitope can be used for localization studies and biochemical experiments with strains grown at the permissive temperature. In addition, the fusion protein is easily purified from cells grown at the permissive temperature using methotrexate-agarose beads because of the high affinity of DHFR for methotrexate.

We will construct a degron fusion to a different kinetochore protein, Cep3p. The integration with the YIp plasmid that produces the degron is shown in Figure 6. To confirm the integration, we will use a PCR primer to the region near the HA sequences at the 3´ end of the degron (primer A) in combination with a primer from within the 5´ end of the *CEP3* gene (Primer B). The latter sequence is not contained within in the *CEP3* fragment in the YIp plasmid but is unique in the chromosomal copy. This assures that the degron is fused to *CEP3*.

STRAIN

7-3 2124   *MATa his3Δ1 ade2-1 ade6-1 hom3-H1 trp1-289 leu2-3,112 ura3-52 bar1::KAN cyh2 can1*

PLASMID

pDB118   An integrating plasmid (YIp) containing *URA3* and the mouse DHFR fused to the 5´ end of *CEP3* and driven by the *CUP1* promoter.

PRIMERS

NDC10-GFP

NDCTAG1
TCAAAATTCATTTGATGGTCTGTTSGTATATCTATCTAACGCGGGTGCTGGAGGAG-
CAATGAG

NDCTAG2
CGGTATCCCTATACGAAACAGTTTAAACTTCGAAGCTCCCCCTCGAGGTCGACG-
GTATC

NDC10-GFP confirmation primers

Primer #1
GGTAACAAGTCGTGGAGAGC

Primer #2
CACGTGTCTTGTAGTTCCCG

Primer #3
TCAGGTAATGGAGCTTTGTCC

Primer #4
CAGGTCTACCAACTCAGTCTTC

Degron Tagging

Primer A
ACCTACCCATACGATGTTCCAG

Primer B
CAGATATGCGGTAACGAGCC

PROCEDURE

Day 1

Streak strain 7-3 for single colonies onto a YPD plate. Incubate 2 days at 30°C.

Day 3

Pick an isolated colony and inoculate 5 ml of YPD liquid medium. Incubate overnight at 30°C.

Day 4

Determine the cell density using a hemocytometer (see Appendix G). Dilute cells to 5×10^6 cells per ml and proceed with a high-efficiency transformation following Techniques and Protocols #1, High-efficiency Transformation of Yeast.

You will be provided with two DNA samples for transformation.

PCR products for GFP tagging: 100 μl containing approximately 8 μg of PCR product. The PCR reactions were previously performed using Techniques and Protocols #13, PCR Protocol for PCR-mediated Gene Disruption using the plasmid pFA6a-GFP(S65T) as a template and primers NDCTAG1 and NDCTAG2. This DNA will be used for four separate transformations.

***Bgl*II-digested pDB118 for degron tagging:** Approximately 1 μg of plasmid pDB118 previously digested with *Bgl*II. This DNA will be used for one transformation.

Be sure to include a negative (no DNA) control in the transformation.

Plate the transformation for GFP tagging onto HC-his plates. Plate the transformation for the degron tagging onto HC-ura plates containing 100 μM CuSO$_4$. Incubate the plates at 23°C for 5 days.

GFP Transformation

Day 7

Reclone the transformants (up to 10) on individual HC-his plates.

Day 10

Pick one colony from each of the transformants to identify GFP fusions. Use Techniques and Protocols #14, Yeast Colony PCR Protocol. Use four different primers, primers #1–#4 in the reaction. Be sure to include cells from strain 7-3 as a control.

Separate the DNA by electrophoresis and identify the GFP fusions. Grow a 5-ml culture of cells in YPD liquid overnight.

Day 11

Inoculate 0.2 ml of the fresh overnight culture into 5 ml of YPD. Grow cells for 4 hours at 30°C to obtain cells that have re-entered the cell cycle. Fix cells in formaldehyde according to Techniques and Protocols #11, Yeast Immunofluorescence, however, reduce the fixation to 15 minutes. Stain cells for both DNA with DAPI and visualize microtubules by anti-tubulin staining using CY3-conjugated secondary antibody.

Degron Transformation

Day 9

Replica-plate the transformants to HC-ura and to HC-ura containing 100 μM CuSO$_4$. Incubate the HC-ura plate at 37°C and the HC-ura containing 100 μM CuSO$_4$ at 23°C.

Day 10

Score the plates in the 37°C incubator and identify five temperature-sensitive colonies. Clone four candidate colonies to a single HC-ura plate containing 100 μM CuSO$_4$. Incubate at 23°C.

Day 14

Pick one colony from each of the transformants to identify degron fusions. Use Techniques and Protocols #14, Yeast Colony PCR Protocol. Use the two primers A and B in the reaction. Be sure to include cells from strain 7-3 as a control. Separate the DNA by electrophoresis and identify one strain containing the degron fusion. Grow a 5-ml culture of cells in HC-ura containing 100 μM CuSO$_4$ overnight.

Day 15

Inoculate 0.2 ml of the fresh overnight culture into 5 ml of HC-ura containing 100 μM CuSO$_4$. Grow cells for 6 hours at 23°C to obtain cells that have re-entered the cell cycle. Remove 2.5 mls of cells, centrifuge, and wash two times with sterile H$_2$O. Incubate the washed cells in HC-ura for 4 hours at 37°C. Return the rest of the culture to 23°C for 4 hours. Fix cells from both cultures in formaldehyde according to Techniques and Protocols #11, Yeast Immunofluorescence; however, reduce the fixation to 15 minutes. Wash the cells twice with 0.1 M potassium phosphate buffer (pH 7.5). Store the cells in a microfuge tube in 1 ml of 0.1 M potassium phosphate buffer (pH 7.5) in the refrigerator overnight.

Day 16

Stain cells for DNA with DAPI according to Techniques and Protocols #11, Yeast Immunofluorescence, beginning at step 5. Visualize microtubules by anti-tubulin staining using CY3-conjugated secondary antibody.

MATERIALS

Day 1   1 YPD plate

Day 3   5 ml of YPD liquid medium

Day 4   Hemocytometer
PCR products for degron tagging (100 µl)
*Bgl*II-digested pDB118 DNA (25 µl)
5 HC-his plates
2 HC-ura plates containing 100 µM CuSO$_4$

Day 7   10 HC-his plates

Day 9   5 ml of liquid YPD medium
1 HC-ura plate
1 HC-ura plate containing 100 µM CuSO$_4$
Sterile velveteen pads

Day 10   1 HC-ura plate containing 100 µM CuSO$_4$
5 ml of liquid YPD medium
Materials for Techniques and Protocols #14, Yeast Colony PCR

Day 11   dH$_2$0
5 ml of liquid YPD medium
Materials for Techniques and Protocols #11, Yeast Immunofluorescence

Day 14   5 ml of liquid HC-ura medium containing 100 µM CuSO$_4$
Materials for Techniques and Protocols #14, Yeast Colony PCR

Day 15   5 ml of liquid HC-ura medium containing 100 µM CuSO$_4$
5 ml of liquid HC-ura medium
37 % formaldehyde
0.1 M phosphate buffer (pH 7.0)

Day 16   Material for Techniques and Protocols #11
CY3-conjugated secondary antibody

REFERENCES

Baudin A., Ozier-Kalogeropoulous O., Deanouel A., LaCroute F., and Cullin C. 1993. A simple and efficient method for direct gene deletion in *Saccharomyces cerevisiae*. *Nucleic Acids Res.* **21**: 3329–3330.

Brachmann C.B., Davies A., Cost G.J., Caputo E., Li J., Hieter P., and Boeke J.D. 1998. Designer deletion strains derived from *Saccharomyces cerevisiae* S288C: A useful set of strains and plasmids for PCR-mediated gene disruption and other applications. *Yeast* **14**: 115–132.

Dohmen R.J., Wu P., and Varshavsky A. 1994. Heat-inducible degron: A method for constructing temperature-sensitive mutants. *Science* **263**: 1273–1276.

Longtine M.S., McKenzie A., Demarini D.J., Shah N.G., Wach A., Brachat A., Philippsen P., and Pringle J.R. 1998. Additional modules for versatile and economical PCR-based gene deletion and modification in *Saccharomyces cerevisiae*. *Yeast* **14**: 953–961.

Lorenz M.C., Muir R.S., Lim E., McElver J., Weber S.C., and Heitman J. 1995. Gene disruption with PCR products in *Saccharomyces cerevisiae*. *Gene* **158**: 113–117.

Rothstein R.J. 1983. One-step gene disruption in yeast. *Methods Enzymol.* **101**: 202–211.

Scherer S. and Davis R.W. 1979. Replacement of chromosome segments with altered DNA sequences constructed in vitro. *Proc. Natl. Acad. Sci.* **76**: 4951–4955.

Sikorski R.S. and Boeke J.D. 1991. *In vitro* mutagenesis and plasmid shuffling: From cloned gene to mutant yeast. *Methods Enzymol.* **194**: 302–328.

Isolation of *ras*2 Suppressors

Suppressor analysis is a powerful genetic approach for the study of genes in a common function or pathway. Suppressor analysis is a secondary approach that is used after an initial mutant with a well-characterized phenotype has been described. The suppressor hunt is designed to allow the identification of additional mutations that restore an aspect of the wild-type phenotype to the initial mutant. The characterization of these suppressor mutations can lead to the identification of new genes with functions related to the initial mutated gene and can suggest interactions between proteins that participate in a common pathway.

RAS1 and *RAS2* specify very similar G-proteins that stimulate adenyl cyclase in yeast. cAMP is essential for growth and for entry into the mitotic cell cycle. The components of the cAMP signal transduction pathway and their biochemical roles are well understood (see Broach 1991), due to a large extent on genetic analysis of the question of cell growth. In this experiment, we follow the approach of Cannon et al. (1986) to isolate suppressors of a *ras2* null mutation.

Under most circumstances, *ras1* or *ras2* single mutants are viable and healthy; one functional RAS protein is sufficient to stimulate adenylyl cyclase. However, *ras2* mutants are defective specifically in growth on nonfermentable carbon sources (such as acetate or glycerol). The reason is that *RAS1* expression is repressed in such growth media. Therefore, a *ras2* null mutation causes a conditional phenotype. We will exploit this phenotype to identify suppressors of *ras2*.

STRAINS

8-6   1784TRP   *MATα RAS2 his4 ura3 leu2 can1*
8-7   AMP141   *MATa ras2-530::LEU2 his4 ura3 leu2 trp1 can1*
8-8   AMP142   *MATα ras2-530::LEU2 his4 ura3 leu2 lys2 can1*

PROCEDURE

Day 1

Instructors have streaked out your *ras2* mutant (strain 8-7 for groups 1, 2, 3, 4; strain 8-8 for groups 5, 6, 7, 8) on a pair of YPD plates.

Construct a patch master on YPD with 30 to 40 isolated colonies, each about 1 cm square. Incubate at 30°C. When you have finished making patches, give one of the plates of single colonies to a group using a mutant hunt strain of the opposite mating type in exchange for a plate of their strain. Refrigerate this plate for future use.

Day 2

Replica-plate the YPD master plate to one YPD plate, then to two YPAc plates. Replica-plate lightly to minimize background growth. Give the instructors one of the replicas for UV treatment (7500 µJ using a Stratalinker). Incubate both plates at 30°C.

Day 6

Papillae (small colonies growing out of the patches) should be visible. Count the number of papillae observed per patch on both the mutagenized and nonmutagenized plates. Pick one papilla from each patch, choosing either the UV-treated plate or the non-UV-treated plate in each case, and patch on a fresh YPD master plate. Pick between 15 and 30 papillae. Pick carefully to avoid contamination with the surrounding cells. (Normally, one would purify the papillae by streaking for single colonies, but we omit that step to save time.) Number each patch and record which plate the papilla came from. Include patches of the parent strain and of the *RAS2* control strain 8-6 on the master plate, and incubate it at 30°C.

Inoculate 5 ml of YPD with the *ras2* strain you obtained from a neighboring group (strain 8-8 for groups 1, 2, 3, 4; strain 8-7 for groups 5, 6, 7, 8). Incubate overnight at 30°C.

Day 7

Pellet the cells in the 5-ml overnight culture started on Day 6 (5 minutes at 2000 rpm in a benchtop centrifuge). Resuspend the culture in 4 ml of YPD. Spread 0.15 ml of the resuspended cells and 0.15 ml of H_2O evenly on an SC–lys–trp plate; place the plate on a level surface and allow to dry.

Replica-plate the YPD master to two YPAc plates, and to two YPD plates and the *ras2* lawn you have prepared on the SC–lys–trp plate. Incubate one YPAc/YPD pair at 30°C, the other at 37°C, and the mating plate at 30°C.

Day 8

Score growth on the 30°C and 37°C plates. Save the 30°C YPD plate for use on Day 12. Choose 12 revertants that you want to analyze in detail. Streak out the crosses of these 12 revertants to the *ras2* lawn on a SC–lys–trp plate. Also streak out the cross of the *ras2* parent to the *ras2* lawn. You should be able to streak out 4–6 crosses per plate. Incubate these plates at 30°C.

Day 10

Pick two isolated colonies from each SC–lys–trp streak and frog to YPAc and YPD plates. Include a colony of the cross of the *ras2* parent to the *ras2* lawn. Incubate at 30°C.

Day 12

Score growth on the YPD and YPAc plates to determine which suppressors are dominant and which are recessive. Go back to the YPD plate you saved on Day 8 and prepare a streak master on YPD, containing parallel streaks of the *ras2* parent and 8–10 recessive suppressor strains (see the streak template in Appendix D). Incubate at 30°C.

Day 13

Replica-plate the streak master from Day 12 to two YPD plates, label the streaks, and exchange one streak plate for one from a group that started their suppressor hunt with a strain of the opposite mating type. Incubate the plates at 30°C.

Day 14

Replica-plate the pair of streak plates perpendicular to each other on a fresh YPD plate. Incubate at 30°C.

Day 15

Replica-plate the crossed streaks from Day 14 to an SC–lys–trp plate. Incubate at 30°C.

Day 16

Pick cells that grew on the SC–lys–trp plate at the streak intersections and use them to prepare a patch master on an SC–lys–trp plate.

Day 17

Replica-plate the Day 16 patch plate (SC–lys–trp) to YPAc and to YPD. Incubate at 30°C.

Day 18

Score the growth of the diploids on the plates from Day 17. What phenotype do you expect from suppressors that fail to complement? Are any of your mutants in the same complementation group as those from your neighbors? How many complementation groups are there?

MATERIALS

Day 1   2 YPD plates with single colonies of a *ras2* strain
1 YPD plate
Toothpicks

Day 2   1 YPD plate
2 YPAc plates
1 Sterile velveteen pad

Day 6   2 YPD plates
Toothpicks
5 ml of YPD

Day 7   3 YPD plates
2 YPAc plates
1 SC–lys–trp plate
4 ml of YPD
1 ml of sterile H_2O
1 Sterile velveteen pad
Toothpicks

Day 8   3 SC–lys–trp plates
Toothpicks

Day 10   Toothpicks
1 YPAc plate
1 YPD plate
1 Sterile 96-well microtiter dish
5 ml of sterile H_2O

Day 12   1 YPD plate
Toothpicks

Day 13   2 YPD plates
1 Sterile velveteen pad

Day 14   1 YPD plate
1 Sterile velveteen pad

Day 15   1 SC–lys–trp plate
1 Sterile velveteen pad

Day 16   1 SC–lys–trp plate

       Toothpicks

Day 17   1 YPAc plate

       1 YPD plate

       1 Sterile velveteen pad

Thanks to Rey Sia for designing the experiment.

REFERENCES

Broach J.R. 1991. RAS genes in *Saccharomyces cerevisiae:* Signal transduction in search of a pathway. *Trends Genet.* **7:** 28–33.

Cannon J.F., Gibbs J.B., and Tatchell K. 1986. Suppressors of the *ras2* mutation of *Saccharomyces cerevisiae. Genetics* **113:** 247–264.

Manipulating Cell Types

Saccharomyces cerevisiae is capable of undergoing sexual reproduction. Haploid cells are able to mate with each other, generating diploid cells. The diploid cells are then capable of undergoing meiosis to regenerate haploid cells. The ability to mate is determined by a cell's mating type. Cells of the same mating type are unable to mate, whereas cells of opposite mating type are able to mate. Mating type is determined by expression of alleles of the *MAT* locus. Cells of one mating type have the MATα allele and are designated α cells; cells of the opposite mating type have the MAT**a** allele and are designated **a** cells. Therefore, α cells are capable of mating with **a** cells, giving rise to diploids that contain both *MAT*α and *MAT***a** alleles. The resulting diploids are designated **a**/α.

The stability of mating type classifies strains into two groups, heterothallic and homothallic. Heterothallic strains have a stable mating type, whereas homothallic haploid strains are capable of changing mating type. Therefore, a colony derived from a homothallic cell will contain both **a** and α cells as well as diploid cells that arose from matings between *a* and α cells.

Spore-to-spore crosses between heterothallic and homothallic strains demonstrated that the distinction between homo- and heterothallism segregates as a single genetic locus designated *HO* (Winge and Roberts 1949; Takahashi et al. 1958). It was shown that a homothallic cell switches from one mating type to the other early in its cell lineage (Strathern and Herskowitz 1979).

A variety of genetic analyses demonstrated the presence of two other loci involved in homothallism (Takahashi 1958; Takano and Oshima 1967). These loci are now designated *HML* and *HMR* and are found with the *MAT* locus on chromosome III (Fig. 1). It has been shown that *HML* and *HMR* contain silent copies of the information expressed from the *MAT* locus (Hicks and Herskowitz 1977; Klar et al. 1979; Strathern et al. 1979). In general, *HML* has the α cell type information and *HMR* has the **a** cell type information. *HO* has been shown to encode a site-specific endonuclease that cuts at a site found specifically within the *MAT* locus. The double-strand break that occurs

CHROMOSOME III

Figure 1. Diagrammatic representation of the silent loci HML and HMR in relation to the *MAT* locus on chromosome III.

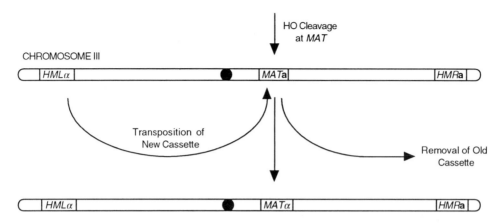

Figure 2. General model of mating type switching by the cassette model. In this case, α information replaces **a** at the *MAT* locus. (Adapted from Herskowitz and Oshima 1981, p. 194.)

allows for recombination between the *MAT* locus and either *HML* or *HMR*. The recombination event, in general, occurs with the silent locus containing information for the opposite mating type from *MAT*. As this recombination event is not reciprocal, the information from the silent locus remains intact while the information at the *MAT* locus changes. These observations led to the cassette model for mating type switching (Fig. 2).

In this experiment we will use (a) *HO* to form diploids and (b) a two-step gene replacement technique to switch the mating type of a strain.

EXPERIMENT IX(a)

USING HO TO GENERATE DIPLOIDS

In this section, a haploid strain will be transformed with a plasmid containing HO to generate diploids. This procedure allows rapid purification of isogenic α and **a** haploid and α/**a** diploid cell types.

STRAINS

9-1 NE2   *MAT*α, *ura3-52, leu2-3,112*
9-2 AAY1017 *MAT*α, *his1*
9-3 AAY1018 *MAT***a**, *his1*

Note: Strains 9-2 and 9-3 are useful for testing mating types, but the genetic background is unknown. Therefore, these strains should never be used for genetic crosses other than mating-type testing.

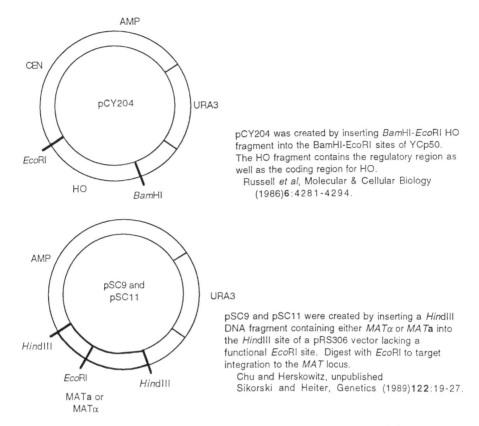

pCY204 was created by inserting *Bam*HI-*Eco*RI HO fragment into the BamHI-EcoRI sites of YCp50. The HO fragment contains the regulatory region as well as the coding region for HO.
 Russell *et al*, Molecular & Cellular Biology (1986)**6**:4281-4294.

pSC9 and pSC11 were created by inserting a *Hind*III DNA fragment containing either *MAT*α or *MAT*a into the *Hind*III site of a pRS306 vector lacking a functional *Eco*RI site. Digest with *Eco*RI to target integration to the *MAT* locus.
 Chu and Herskowitz, unpublished
 Sikorski and Heiter, Genetics (1989)**122**:19-27.

Figure 3. Plasmids for Experiment IX, Sections (a) and (b).

PLASMID

pCY204 *HO* in YCp50. Selectable marker is *URA3* (Fig. 3).

PROCEDURE

Day 1

Streak strain 9-1 for single colonies on a YPD plate and incubate at 30°C.

Day 3

In the morning, inoculate 5 ml of YPD with a colony of strain 9-1 and incubate at 30°C. Store the YPD plate from above at 4°C for use on Day 13.

Day 4

Transform strain 9-1 following Techniques and Protocols #1, High-efficiency Transformation of Yeast, with uncut plasmid pCY204 (from your instructor).

Remember to perfrom a no-DNA control transformation as well. Plate one 200 µl aliquot of transformed cells on an SC–ura plate and incubate at 30°C. Do the same for the no-DNA control.

Day 7

Streak 4 transformants for single colonies on SC–ura and grow at 30°C.

Day 9

Inoculate 5 ml of YPD with a single colony from each individual transformant and grow at 30°C.

Day 10

Determine cell density using a hemocytometer. Dilute cells appropriately in sterile distilled H_2O and plate 10^6 and 10^5 cells in 100-µl volumes on 5-FOA plates. This will select against the *URA3* gene so that cells that have lost pCY204 can grow. Store the remaining YPD cultures at 4°C for use on Day 12.

Day 12

Make a patch master plate by patching nine 5-FOA^R colonies of each transformant onto a YPD plate (36 total). Also patch the parent strain 9-1. In addition, spot 3 µl of each of the four intermediate cultures from Day 10 onto the patch master plate. Incubate the YPD plate at 30°C. Inoculate two 5-ml YPD cultures with strain 9-2 and strain 9-3. Grow at 30°C with agitation.

Day 13

On an SD plate, spread 150 µl of strain 9-2 and 150 µl of sterile H_2O and allow to dry. Repeat this process with another SD plate using strain 9-3. Replica-plate the patch master plate onto the SD plates with each lawn. Use a fresh velvet for each plate to pre-vent contamination between plates. Also replica-plate the patch master plate onto a sporulation plate, and onto an SC–ura plate.

Day 15

Score mating ability and growth on the SC–ura plate.

Day 17

Score sporulation ability.

MATERIALS

Day 1   1 YPD plate

Day 3   1 Culture tube containing 5 ml of YPD

Day 4   Materials for Techniques and Protocols #1, High-efficiency Trans-
               formation of Yeast
               Uncut plasmid pCY204
               Erlenmeyer flask containing 50 ml of YPD
               2 SC–ura plates

Day 7   1 SC–ura plate

Day 9   4 Culture tubes containing 5 ml of YPD

Day 10   Sterile dH$_2$O
               8 5-FOA plates

Day 12   1 YPD plate
               2 Culture tubes containing 5 ml of YPD

Day 13   2 Sterile velveteen pads
               2 SD plates
               1 Sporulation plate
               1 SC–ura plate

EXPERIMENT IX(b)

MATING-TYPE SWITCHING BY TWO-STEP GENE REPLACEMENT

Two-step gene replacement allows the replacement of one allele of a gene with another allele (see also Experiment VII). This technique is generally used to replace a wild-type allele with a mutant allele that has no selectable phenotype. The first step of this procedure involves integrating a plasmid sequence containing a selectable marker and the gene of interest. This type of integration leads to a duplication of the gene of interest, with the two genes separated by the plasmid sequence (Fig. 4A). The second step is loss of the plasmid sequence by homologous recombination between the duplicated regions. Loss of the plasmid is most easily accomplished if the marker used to select for integration can be selected against, for example, by selecting against *URA3* with

5-fluoro-orotic acid (5-FOA). Depending on where the recombination event occurs, the remaining copy will carry one allele or the other (Fig. 4B). In this experiment, we will use two-step gene replacement to switch mating types (S. Chu and I. Herskowitz, unpubl.).

STRAINS

9-4 YSC006 *MATα ura3 ade2-1 trp1-1 can1-100 leu2-3,112 his3-11,15[psi⁺]GAL⁺*

9-5 YSC005 *MAT**a** ura3 ade2-1 trp1-1 can1-100 leu2-3,112 his3-11,15[psi⁺]GAL⁺*

PLASMIDS

pSC9   *MATα* in pRS306 selectable marker is *URA3* (Fig. 4A)

pSC11   *MAT**a*** in pRS306 selectable marker is *URA3* (Fig. 4A)

PROCEDURE

Day 1

Streak strain 9-4 or 9-5 for single colonies onto YPD plates and incubate at 30°C. Groups 1, 2, 3, and 4 will use strain 9-4 and groups 5, 6, 7, and 8 will use strain 9-5.

Day 3

In the morning, inoculate 5 ml of YPD with a single colony and incubate at 30°C. Store the YPD plate from above at 4°C for use on Day 13.

Day 4

Transform strain 9-4 or 9-5 following Techniques & Protocols #1, High-efficency Transformation of Yeast. Obtain digested plasmid DNA from instructor for transformation. Strain 9-4 will be transformed with pSC11 and strain 9-5 with pSC9. Remember to perform a no-DNA control transformation as well. Plate one 200-μl aliquot of transformed cells on an SC–ura plate and incubate at 30°C. Do the same for the no-DNA control.

Day 7

Streak 4 transformants for single colonies on SC–ura and grow at 30°C.

Day 9

Inoculate 5 ml of YPD with a single colony from each individual transformant and grow at 30°C.

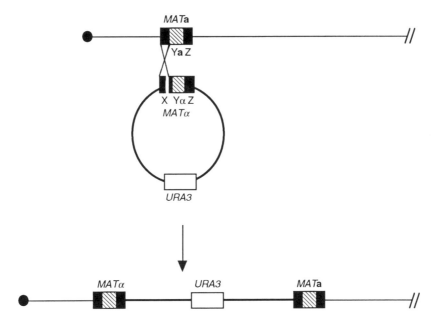

Figure 4A. Two-step gene replacement.

Step 1: Integration of plasmid by homologous recombination at duplicated regions, either X or Z. Unlike one-step gene replacements, the entire plasmid is integrated. *Note:* The relative order of the two alleles is determined by the position of the crossover. In this figure, sequences determining mating-type occur 3′ of the crossover event.

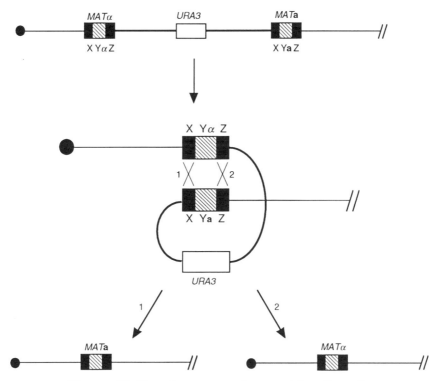

Figure 4B. Two-step gene replacement (continued).

Step 2: Selection against the integration (in this case, selecting against *URA3* by plating on 5-FOA) allows for "looping out" by homologous recombination between the duplicated regions of the *MAT* locus (X or Z). Depending where recombination occurs (X or Z), the mating type will either be switched to *MAT*α or reverted to the parent allele (*MAT*a).

Day 10

Determine cell density using a hemocytometer (see Appendix G). Dilute cells appropriately in sterile distilled H_2O and plate 10^6 and 10^5 cells in 100-μl volumes on 5-FOA plates. This will select against the *URA3* gene that should be inserted between the duplicated regions of the *MAT* locus. 5-FOA[R] arises through "looping out" of the *URA3* gene by homologous recombination. Store the remaining YPD cultures at 4°C for use on Day 14.

Day 13

Make a patch master plate by patching nine 5-FOA[R] colonies of each transformant onto a YPD plate (36 total). Also patch the parent strains 9-4 and 9-5. (Obtain one parent strain from the plate stored at 4°C on Day 3, and the other parent from a neighboring group.) In addition, spot 10 μl of each of the four intermediate cultures from Day 8 onto the patch master plate. Incubate the YPD plate at 30°C.

Day 14

On an SD plate, spread 100 μl of strain 9-2 (from Experiment IX[a], Day 13) and allow to dry. Repeat this process with another SD plate using strain 9-3. Replica-plate the patch master plate onto the SD plates with each lawn. Use a fresh velvet for each plate to prevent contamination between plates. Also replica-plate the patch master plate onto a sporulation plate and onto an SC–ura plate.

Day 16

Score mating ability and growth on the SC–ura plate.

Day 18

Score sporulation ability.

MATERIALS

Day 1   1 YPD plate

Day 3   Culture tube containing 5 ml of YPD

Day 4   Materials for Techniques and Protocols #1, High-efficiency Transformation of Yeast
Digested plasmid pSC9 and pSC11 DNA
Erlenmeyer flask containing 50 ml of YPD
2 SC–ura plates

Day 7   1 SC–ura plates

Day 9   4 Culture tubes containing 5 ml of YPD

Day 10   Sterile distilled H_2O
             8 5-FOA plates

Day 13   1 YPD plate

Day 14   1 Sporulation plate
             1 SC–ura plate
             2 SD plates
             2 Sterile velveteen pads

The method of switching mating-type by two-step gene replacement was developed by Shelley Chu and Ira Herskowitz. Thanks to them for providing the methodology and materials for this procedure. Thanks to Dana Davis for writing up the chapter.

REFERENCES

Herskowitz I. and Oshima Y. 1981. Control of cell type in *Saccharomyces cerevisiae*: Mating type and mating-type interconversion. In *The molecular and cellular biology of the yeast* Saccharomyces: *Life cycle and inheritance* (ed. J.N. Strathern et al.), pp. 181-209. Cold Spring Harbor Laboratory, Cold Spring Harbor, New York.

Hicks J.B. and Herskowitz I. 1977. Interconversion of yeast mating types. II. Restoration of mating ability to sterile mutants in homothallic and heterothallic strains. *Genetics* **85**: 373.

Klar A.J.S., Fogel S., and Radin D.N. 1979. Switching of a mating-type **a** mutant allele in budding yeast *Saccharomyces cerevisiae*. *Genetics* **92**: 759.

Russell D.W., Jensen R., Zoller M.J., Burke J., Errede B., Smith M., and Herskowitz I. 1986. Structure of the *Saccharomyces cerevisiae HO* gene and analysis of its upstream regulatory region. *Mol. Cell. Biol.* **6**: 4281–4294.

Sikorski R.S. and Hieter P. 1989. A system of shuttle vectors and yeast host strains designed for efficient manipulation of DNA in *Saccharomyces cerevisiae*. *Genetics* **122**: 19–27.

Strathern J.N. and Herskowitz I. 1979. Asymmetry and directionality in production of new cell types during clonal growth: The switching pattern of homothallic yeast. *Cell* **17**: 371–381.

Strathern J.N., Blair L.C., and Herskowitz I. 1979. Healing of *mat* mutations and control of mating type interconversion by the mating type locus in *Saccharomyces cerevisiae*. *Proc. Natl. Acad. Sci.* **76**: 3425.

Takahashi T. 1958. Complementary genes controlling homothallism in *Saccharomyces*. *Genetics* **43**: 705.

Takahashi T., Saito H., and Ikeda Y. 1958. Heterothallic behavior of a homothallic strain in *Saccharomyces cerevisiae*. *Genetics* **43**: 249–260.

Takano I. and Oshima. Y. 1967. An allele specific and a complementary determinant controlling homothallism in *Saccharomyces oviformis*. *Genetics* **57**: 875.

Winge Ø. and Roberts C. 1949. A gene for diploidization in yeast. *C.R. Trav. Lab. Carlsberg Ser. Physiol.* **24**: 341–346.

Isolating Mutants by Insertional Mutagenesis

There are two kinds of microtubules in yeast that have different functions in the life cycle. The nuclear microtubules are the major components of the intranuclear mitotic spindle and are required for chromosome separation. The cytoplasmic microtubules are required for nuclear positioning between the daughter cells and for nuclear fusion during mating. Mutants lacking cytoplasmic microtubules are viable demonstrating that their functions are dispensable. Cells lacking microtubules are inviable suggesting that the essential function of the microtubules is chromosome segregation.

Benomyl is a benzimidazole drug that is a potent inhibitor of microtubule assembly. At high concentration, the drug arrests cells at mitosis and at lower concentration cells delay prior to anaphase in the cell cycle. The sublethal doses of benomyl are especially useful for genetic screens and two classes of mutants have been identified. One class of benomyl sensitive mutants includes genes that affect the essential microtubule functions. The other class of mutants affects cell cycle regulation and the ability to arrest in the cell cycle in response to microtubule dysfunction.

INSERTIONAL MUTAGENESIS

In this experiment we will screen for benomyl sensitive mutants using insertional mutagenesis. A library of yeast genes was constructed in a bacterial vector and the library was mutagenized with a transposon in *E. coli* (Burns et al. 1994; Ross-Macdonald et al. 1999). A modified Tn3 transposon contains the *LEU2* gene from yeast and the *lacZ* gene from *E. coli* subcloned between the 38-bp inverted repeats that marks the end of the transposon. The transposon lacks all functions necessary for transposition (they are supplied in *trans*) and is referred to as a mini-transposon or mTn3-*lacZ/LEU2*. The *lacZ* gene is adjacent to one of the 38-bp repeats and lacks the codon for initiating translation. Therefore, it is possible that some of the transpositions (approximately 1/6) are in frame *lacZ* fusions. The modified version of the *lacZ* gene is sometimes referred to as *lacZ′*. The transposon mutagenesis is accomplished by a two-step bacterial mating. The strain containing the library is successively mated to two strains that supply mTn3, the transposase and resolvase functions in *trans*. The result is that the library is recovered as a large number of plasmids (greater than 10^5), each of which contains a different transposition event. The plasmid DNA is recovered and digested with the

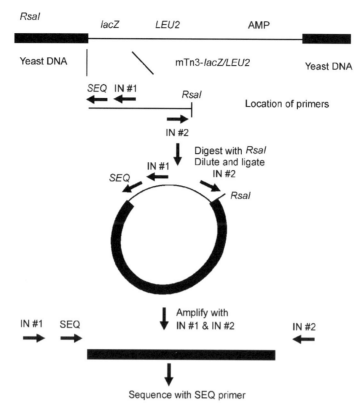

Figure 1. Inverse PCR.

8-bp restriction enzyme *NotI* which excises the yeast inserts containing the mTn3. The DNA is transformed into a *leu2* yeast strain to produce one step gene replacements. In this case, the "replacement" fragment is a randomly mutagenized region of the yeast genome.

There are two important characteristics of transposon mutagenesis. The first is that the mutagen (the transposon) generates insertion mutations. Insertions into essential genes are generally lethal events and are not recovered if the transformation is done into haploid cells. This introduces a significant bias in the experiment as we expect to primarily recover mutations in nonessential genes. The second characteristic is that the insertion marks the mutated gene providing a powerful and effective tool for molecular cloning and identification. We will make use of the highly efficient recombination system in *Saccharomyces cerevisiae* to construct mutants. We will exploit the fully sequenced genome to identify the genes.

There are two ways to identify the mutant. The first is to rescue the transposon and adjacent sequences as a plasmid. In this experiment, we will employ the second method. We will recover the flanking sequences by "inverse PCR." The procedure is illustrated in Figure 1. The thin line represents mTN3 *lacZ/LEU2* transposon DNA and the thick line represents yeast DNA. Primers from the sequences at the *lacZ end* of the transposon are shown schematically as arrows. The restriction enzyme site for *RsaI*, which recognizes a four base pair sequence (GTAC), is shown. One site is adjacent to the IN #2 primer used for inverse PCR. The other RsaI site some unknown distance within the yeast DNA. The distance is likely to be small because of the high density of *RsaI* sites in the genome. Yeast DNA is digested with *RsaI*, diluted and ligated. The dilution promotes intramolecular ligation and the formation of circles. The IN #1 and IN #2 primers are used to amplify a fragment that contains the yeast DNA sequences from the *lacZ´*: end of the transposon to the flanking RsaI site. The sequence of the DNA is determined using the SEQ primer to sequence the DNA fragment.

We will employ two different strategies for isolating mutants in this experiment. Half of the class will screen for benomyl sensitivity. A wild-type culture of cells will be mutagenized and the surviving cells will be plated onto HC-leu plates to select transformants. Colonies will be replica-plated to HC-leu plates containing benomyl and screened for benomyl sensitive mutants. The remainder of the class will select for benomyl sensitivity by suppressing the lethality of excess Mps1p derived from a P_{GAL10}-*MPS1* fusion gene. The *MPS1* gene encodes a protein kinase that is a component of the spindle checkpoint and excess expression for the *GAL10* promoter causes cells to arrest in mitosis. We will select for genes on the *MPS1*-dependent signal transduction pathway. A strain containing P_{GAL10}-*MPS1* and *cdc28^{Y19F}* will be mutagenized and plated onto HC-leu plates containing galactose. The *cdc28^{Y19F}* mutation eliminates mitotic regulation of Cdc28p by reversible phosphorylation and increases the efficiency of the cell cycle arrest induced by excess Mps1p.

STRAINS

10-1	2124	*MATa his3 ade2 ade6 cyh2 can1 hom3 trp1 leu2 ura3 bar1::KAN*
10-2	2194-12-2	*MATα his3 leu2 trp1 ura3 cdc28F19::TRP1 GAL-MPS1::URA3 can1*

PRIMERS

IN #1	TAAGTTGGGTAACGGCCCAGGGTTTTC
IN #2	TGTTGCCACTCGCTTTTAATG
SEQ	CCCCCTTAACGTTCCACT

PROCEDURE

Day 1

Streak strains for single colonies onto a YPD plate. Incubate for two days at 30°C.

Day 3

Pick an isolated colony and inoculate 5 ml of YPD liquid medium. Incubate overnight at 30°C

Day 4

Determine the cell density using a hemocytometer (see Appendix G). Dilute cells to 5×10^6 cells per ml and proceed with a high-efficiency transformation following Techniques and Protocols #1, High-efficiency Transformation of Yeast. There will be a slight modification as described below.

Harvest cells and transform with 9 µg of mTn3-*lacZ/LEU2* genomic library DNA previously digested with *NotI*. You will have a total of 18 µg of DNA in a total volume of 180 µl. Use 10 µl of DNA for each transformation. Perform 10 transformations for each strain, one with no DNA as a control and nine with 1 µg of DNA.

STRAIN 10-1: At step 16, resuspend the cells in 600 µl of sterile water and plate 200 µl of the transformants onto HC-leu plates. Incubate for 3 days at 30°C.

STRAIN 10-2: At step 16 resuspend the cells in 200 µl of distilled water and plate the 200 µl of cells from each transformation onto a single HC-leu plate containing galactose. Incubate for 4 days at 30°C

Strain 10-1 Transformation

Day 9

Determine the number of Leu⁺ colonies from the strain 10-1 transformation. Replica-plate the transformants to HC-leu and HC-leu containing 15 µg/ml benomyl. Incubate at 30°C overnight.

Day 10

Screen for benomyl sensitivity. Determine the proportion of benomyl-sensitive colonies among the transformants. Re-clone six benomyl-sensitive colonies from HC-leu plates to new HC-leu plates. Incubate for two days at 30°C.

Day 12

Pick isolated colonies and grow 5-ml cultures in YPD overnight at 30°C.

Strain 10-2 Transformation

Day 8

Replica-plate to HC-leu and to HC-leu plates containing 15 µg/ml of benomyl. Incubate at 30°C.

Day 10

Identify benomyl sensitive colonies and recover the same colonies from HC-leu plates. Re-clone on HC-leu plates at 30°C.

Day 12

Pick isolated colonies and grow no more than six 5-ml cultures in YPD overnight at 30°C.

Stains 10-1 and 10-2 Transformations

Day 13

Prepare genomic DNA from the six cultures using Techniques and Protocols #3D, *Yeast Genomic DNA: Glass Bead Preparation*, and resuspend the final pellet in 100 μl of TE. Digest 15 μl of each DNA with 5–10 U of *RsaI* in a total volume of 100 μl. Add 1 μg of RNase A and incubate for several hours at 37°C. Inactivate the *RsaI* by heating to 65°C for 20 minutes. Remove 10 μl and dilute to 100 μl in 1 x T4 DNA ligase buffer (with ATP). Add 100 U of T4 DNA ligase and incubate overnight at room temperature.

Day 14

Amplify the DNA adjacent to the Tn3 transposon by PCR. To 5 μl of ligated DNA add:

5 μl of 10x *Taq* Polymerase buffer lacking Mg^{++}
1 μl of 25 μM Primer #1
1 μl of 25 μM Primer #2
1 μl of 10 mM dNTPs
32 μl of distilled H$_2$0
1 μl of *Taq* Polymerase (2.5 U)

It is critical to use a "hot start" protocol for successful amplification of isolated genomic DNA. We will use "Mg HotBeads" from Lumitekk. These are wax beads containing the MgCl$_2$ needed for the PCR reactions. Add 1 wax bead to each PCR reaction. The wax will melt at 68°C to release MgCl$_2$. Amplify with:

5 minutes at 95°C, followed by 35 cycles of:

1 minute at 94°C,
1 minute at 65°C,
2.5 minutes at 72°C.

Conclude the final cycle with 7 minutes at 72°C.

Purify the amplified DNA away from protein, oligonucleotides, and dNTPs (a Wizard PCR Preps kit will be provided in the course), and submit for DNA sequencing.

Day 15

Using a computer that is connected to the Internet, submit the DNA sequences for comparison to the yeast genome database and discover where your inserts landed!

MATERIALS

Day 1   1 YPD plate

Day 3   1 Culture tube containing 5 ml of YPD

Day 4   Erlenmeyer flask containing 50 ml of YPD
               Materials for Techniques and Protocols #1, High-efficiency Trans-
                  formation of Yeast
               0.5 mg of carrier DNA
               9 μg of *NotI* cleaved mTN3-*lacZ/LEU2* genomic library DNA
               30 HC-leu plates for strain 10-1 transformation
               10 HC-leu plates for strain 10-2 transformation

Day 7   10 HC-leu plates containing galactose (10-2 transformation)

Day 9   30 HC-leu plates and 30 HC-leu plates containing 15 μg/ml benomyl
               (10-1 transformation)
               30 Sterile velveteen pads (10-1 transformation)
               10 HC-leu plates containing galactose (10-2 transformation)

Day 10   6 HC-leu plates (strain 10-1 transformation)

Day 12   6 culture tubes containing 5 ml of liquid YPD medium

Day 13   *RsaI*
               60 μl of 10x restriction enzyme buffer
               5 μl of RNaseA (5 mg/ml)
               60 μl of T4 DNA ligase buffer +ATP
               T4 DNA ligase

Day 14   Mg HotBeads
               50 μl of 10x *Taq* polymerase buffer
               10 μl of 25 M primer #1
               10 μl of primer #2
               10 μl of 10 mM dNTPs

dH$_2$O
Taq polymerase
Wizard PCR preps DNA purification system (Promega)
ABI Dye terminator DNA sequencing kit with mTn3 sequencing primer
(Applied Biosystems Inc.)

Day 15   Networked computer

REFERENCES

Burns N., Grimwade B., Ross-Macdonald P.B., Choi E.-Y., Finberg K., Roeder G.S., and Snyder M. 1994.
    Large-scale characterization of gene expression, protein localization and gene disruption in
    Saccharomyces cerevisiae. Genes Dev. **8**: 1087–1105.
Ross-Macdonald P., Coelho P.S., Roemer T., Agarwal S., Kumar A., Jansen R., Cheung K.H., Sheehan A.,
    Symoniatis D., Umansky L., Heidtman M., Nelson F.K., Iwasaki H., Hager K., Gerstein M., Miller
    P., Roeder G.S., and Snyder M. 1999. Large-scale analysis of the yeast genome by transposon tag-
    ging and gene disruption. *Nature* **402**: 413–418.

Two-hybrid Protein Interaction Method

Most biological processes are mediated by protein-protein interactions, and a wealth of biochemical assays have been developed to detect such interactions. The two-hybrid system is a yeast-based genetic assay that provides a simple and sensitive means to detect potential interactions between two proteins. It is based on the finding that certain eukaryotic transcription activators are modular. For example, Gal4p, a positive regulator of the *GAL* pathway, consists of a site-specific DNA-binding domain and an acidic transcription activation domain (Keegan et al. 1986). The DNA-binding domain binds to an upstream activating sequence (UAS) found in the promoters of GAL genes. The transcription activation domain interacts with other components of the transcription apparatus to initiate transcription. These two domains are separable and can function as independent units. Fields and Song (1989) used this feature of Gal4p to create a system in which two hybrid proteins are created: One protein is fused to the Gal4p DNA-binding domain, and another is fused to the Gal4p transcription activation domain. The binding and activation domains themselves do not interact with each other and, when expressed together in a yeast cell, are unable to activate transcription at a *GAL* promoter. However, if the sequences fused to the binding and activation domains are able to interact with each other, then the binding and activation domains are brought together on the DNA and a functional transcriptional activator is reconstituted.

In the simplest case, the two-hybrid system is used to test for interactions between two known proteins; this is often referred to as a "directed" two-hybrid. A variation of the directed two-hybrid is to test the ability of mutant forms (typically deletions) of the proteins of interest to interact, potentially identifying domains required for the interaction. Another use of the two-hybrid system is to identify interacting proteins from a random genomic or cDNA library. In this case, the protein of interest is fused to the DNA-binding domain, and random genomic or cDNA fragments are fused to the activation domain. In the experiment described here, a variation of the original yeast two-hybrid system developed by Elledge (Harper et al. 1993) will be used to examine the interaction between two proteins, Tub1p and Alf1p. Tub1p is the major α-tubulin in yeast; Alf1p is involved in the formation of tubulin and binds to α-tubulin in both two-hybrid and immunoprecipitation assays (Tian et al. 1997; Feierbach and Stearns 1999).

The activity of the hybrid transcriptional activator is assayed using constructs that have a GAL promoter driving one of a variety of reporter genes. In the version used in this experiment, the two reporter genes are *HIS3*, a yeast gene involved in histidine biosynthesis, and *lacZ*, the bacterial gene for β-galactosidase. The *HIS3* reporter allows selection of His⁺ cells on medium lacking histidine. The *lacZ* reporter allows visual screening of cells expressing β-galactosidase, as visualized with the chromogenic substrate X-gal. The promoters driving these reporters differ in the number and affinity of upstream binding sites for the DNA-binding hybrid and in the position of these sites relative to the transcription start-point. These differences affect the strength of the protein interaction the reporters can detect; activation of the *lacZ* reporter is the more rigorous test for function.

Several requirements must be fulfilled for the two-hybrid method to succeed. First, both hybrid proteins must be expressed. In some versions of the two-hybrid vectors, an epitope tag is included in the construct, allowing expression to be verified by Western blotting. Second, neither hybrid protein should be able to activate transcription by itself. When identifying interactors from a library, there are several possibilities for false positives, not all of which are understood mechanistically. To help to identify such false positives, plasmids that are recovered from the library are subjected to additional tests: first, for self-activation of the reporter genes, then for specificity of interaction with protein of interest, as compared to several standard test proteins (p53 and lamin are typically used). As with all genetic methods, interactions detected in the two-hybrid system should be confirmed by biological and/or biochemical experiments.

STRAINS

11-1   Y190   *MATa gal4 gal80 his3-Δ200 trp1-901 ade2-101 leu2-3,112 lys2:GAL-HIS3:LYS2 ura3-52:GAL-lacZ:URA3 Cyhʳ*

PLASMIDS

pAS1   *GAL4* DNA-binding domain vector (*TRP1* is selectable marker)
pACTII   *GAL4* activation domain vector (*LEU2* is selectable marker)
pTS800   *TUB1* in pAS1 (DNA-binding domain fusion)
pTS861   *ALF1* in pACTII (*GAL4* activation domain fusion)

PROCEDURE

SAFETY NOTE

The temperature of liquid nitrogen (N₂) is –185°C. Use with extreme caution. Always wear thermal gloves and face protection to prevent burns.

Day 1

Streak strain 11-1 for single colonies on YPD in preparation for transformation. Incubate at 30°C.

Day 3

In the morning, inoculate 5 ml of YPD with a single colony of strain 11-1 to make a preculture. Incubate at 30°C with agitation. In the evening, determine the cell concentration using a hemocytometer. Assuming an approximate generation time of 100 minutes, calculate the volume of preculture to be added to 50 ml of YPD in order for the culture to reach a concentration of 2×10^7 cells/ml by 10 a.m. on Day 4.

Day 4

Harvest the cells by centrifugation at 2000 rpm in a clinical centrifuge and follow the protocol for LiAc transformation given in Techniques and Protocols #1, High-efficiency Transformation of Yeast. Transform strain 11-1 with the plasmid combinations listed below. Note that each transformation includes *two* plasmids, and that the selection is for the markers on each of the two plasmids. Incubate at 30°C on the indicated plate.

Plasmids	Selective Medium
pTS800 + pTS861	SC-leu-trp
pTS800 + pACTII	SC-leu-trp
pAS1 + pTS861	SC-leu-trp
pAS1 + pACTII	SC-leu-trp

Day 6

(i) Streak two colonies from each of the four transformations onto SD+ ade+25 mM 3-aminotriazole (3-AT) plates to test for activation of the *HIS3* gene. Incubate overnight at 30°C. 3-Aminotriazole is a competitive inhibitor of the His3p enzyme and is necessary to make the His+ selection more stringent.

(ii) Patch two colonies of each transformation onto SC-leu-trp plates for the X-gal filter lift assay. Make the patches large, using most of the plate for the 8 strains. Incubate overnight at 30°C.

Day 7

Check for growth on plates containing 3-AT. Growth indicates activation of the *HIS3* gene. Depending on the transformant tested, this activation can indicate an interaction between two proteins, or self-activation.

Day 8

(i) Check the 3-AT plates again for growth.

(ii) X-gal Filter-Lift Assay:

1. Using a pencil or ballpoint pen, mark a 3MW paper filter so that you can determine orientation. The filter should be cut into the shape of a circle that will fit into a petri dish (precut filters can be purchased).

2. Carefully place the filter on the SC-leu-trp plate, gently pressing down to remove any bubbles.

3. Using forceps, carefully remove the filter from the plate and immerse in liquid N_2 for one minute to permeabilize the cells. *BE CAREFUL: Wear thermal gloves and eye protection to prevent burns.*

4. Remove the filter, let it thaw completely, then repeat the liquid N_2 immersion/thawing process two more times.

5. After the filter thaws the third time, gently place it, with the cells facing upward, onto an X-gal plate. Incubate at 30°C and examine the plate every hour over the course of the day, noting any blue color development of the cells.

MATERIALS

Day 1   1 YPD plate

Day 3   5 ml of YPD in test tube
               50 ml of YPD in flask

Day 4   Materials for Techniques and Protocols #1
               4 SC-leu-trp plates

Day 6   2 SD+ade+25 mM 3-AT plates
               1 SC-leu-trp plate

Day 8   3MW filters
               Liquid N_2
               1 X-Gal plate

REFERENCES

Feierbach B., Nogales E., Downing K.H., and Stearns T. 1999. Alf1p, a CLIP-170 domain-containing protein, is functionally and physically associated with α-tubulin. *J. Cell Biol.* **144:** 113–124.

Fields S. and Song O. 1989. A novel genetic system to detect protein-protein interactions. *Nature* **340:** 245–246.

Harper J.W., Adami G.R., Wei N., Keyomarski K., and Elledge S.J. 1993. The p21 Cdk-interacting protein Cip1 is a potent inhibitor of G1 cyclin-dependent kinases. *Cell* **75:** 805–816.

Keegan L., Gill G., and Ptashne M. 1986. Separation of DNA binding from the transcription-activating function of a eukaryotic regulatory protein. *Science* **231:** 699–704.

Tian G., Lewis S.A., Feierbach B., Stearns T., Rommelaere H., Ampe,C., and Cowan N.J. 1997. Tubulin subunits exist in an activated conformational state generated and maintained by protein cofactors. *J. Cell Biol.* **138:** 821–832.

High-efficiency Transformation of Yeast

PROCEDURE

This protocol was adapted from Gietz and Schiestl (1995).

1. Inoculate 5 ml of liquid YPAD or 10 ml of SC and incubate with shaking overnight at 30°C.

2. Count overnight culture and inoculate 50 ml of YPAD to a cell density of 5 × 10^6/ml of culture.

3. Incubate the culture at 30°C on a shaker at 200 rpm until it is at 2 × 10^7 cells/ml. This typically takes 3–5 hours. This culture will give sufficient cells for 10 transformations.

   Notes:
   i. It is important to allow the cells to complete at least 2 divisions.
   ii. Transformation efficiency remains constant for 3–4 cell divisions.

4. Harvest the culture in a sterile 50-ml centrifuge tube at 3000*g* (2500 rpm) for 5 minutes.

5. Pour off the medium, resuspend the cells in 25 ml of sterile H_2O, and centrifuge again.

6. Pour off the H_2O, resuspend the cells in 1.0 ml of 100 mM lithium acetate (LiAc), and transfer the suspension to a sterile 1.5-ml microfuge tube.

7. Pellet the cells at top speed for 5 seconds and remove the LiAc with a micropipette.

8. Resuspend the cells to a final volume of 500 µl (2 × 10^9 cells/ml), which is about 400 µl of 100 mM LiAc.

   Note: If the cell titer of the culture is greater than 2 × 10^7 cells/m, the volume of the LiAc should be increased to maintain the titer of this suspension at 2 × 10^9 cells/ml. If the titer of the culture is less than 2 × 10^7 cells/ml, then decrease the amount of LiAc.

9. Boil a 1.0-ml sample of single-stranded carrier DNA for 5 minutes and quickly chill in ice water.

   Note: It is not necessary or desirable to boil the carrier DNA every time. Keep a small aliquot in your own freezer box and boil after 3–4 freeze-thaws. But keep on ice when out.

10. Vortex the cell suspension and pipette 50-µl samples into labeled microfuge tubes. Pellet the cells and remove the LiAc with a micropipette.

11. The basic "transformation mix" consists of the following ingredients; carefully add them *in the order listed:*

> 240 µl of PEG (50% w/v)
>
> 36 µl of 1.0 M LiAc
>
> 25 µl of single-stranded carrier DNA (2.0 mg/ml)
>
> 50 µl of H_2O and plasmid DNA (0.1–10 µg)

Note: The order is important here! The PEG, which shields the cells from the detrimental effects of the high concentration of LiAc, should go in first.

12. Vortex each tube vigorously until the cell pellet has been completely mixed. This usually takes about 1 minute.

13. Incubate for 30 minutes at 30°C.

14. Heat shock for 20–25 minutes in a water bath at 42°C.

Note: The optimum time can vary for different yeast strains. Please test this if you need high efficiency from your transformations.

15. Microfuge at 6000–8000 rpm for 15 seconds and remove the transformation mix with a micropipette.

16. Pipette 0.2–1.0 ml of sterile H_2O into each tube and resuspend the pellet by pipetting it up and down gently.

Note: Be as gentle as possible at this step if high efficiency is important.

17. Plate from 200-µl aliquots of the transformation mix onto selective plates.

MATERIALS AND SOLUTIONS

Polyethylene glycol (PEG; 50% w/v) (MW 3350; Sigma P 3640)

Make up to 50% (w/v) with H_2O and filter-sterilize with a 0.45-µm filter unit (Nalgene). Alternatively, the PEG solution can be autoclaved, but care must be taken to ensure that the PEG solution is at the proper concentration. In addition, it is important to store the PEG in a tightly capped container to prevent evaporation of H_2O and a subsequent increase in PEG concentration. Small variations above or below the PEG concentration optimum in the transformation reaction, which is 33% (w/v), can reduce the production of transformants.

Single-stranded carrier DNA (2 mg/ml)

High-molecular-weight DNA (Deoxyribonucleic acid Sodium Salt Type III from Salmon Testes; Sigma D 1626)

TE buffer (pH 8.0)

10 mM Tris-HCl (pH 8.0)

1.0 mM EDTA

1. Weigh out 200 mg of the DNA into 100 ml of TE buffer. Disperse the DNA into solution by drawing it up and down repeatedly in a 10-ml pipette. Mix vigorously on a magnetic stirrer for 2–3 hours or until fully dissolved. Alternatively, leave the covered solution mixing at this stage overnight in a cold room.

2. Aliquot the DNA (100 μl is typically convenient), and store at –20°C.

3. Prior to use, an aliquot should be placed in a boiling water bath for at least 5 minutes and quickly cooled in an ice-water slurry.

TIPS:

i. Carrier DNA can be frozen after boiling and used 3 or 4 times. If transformation efficiencies begin to decrease with a batch of boiled carrier DNA, it should be boiled again or a new aliquot used.

ii. The lower concentration of carrier DNA (2 mg/ml) in this protocol eases handling and gives more reproducible results.

iii. In previous protocol versions, a phenol:chloroform extraction was used to ensure maximal transformation efficiencies. This extraction may not be necessary if the DNA is of high enough quality. Test your carrier DNA to determine if extraction is necessary.

1.0 M Lithium acetate stock solution (LiAc)

Prepare as a 1.0 M stock in distilled deionized H_2O; filter-sterilize. There is no need to titrate this solution, but the final pH should be between 8.4 and 8.9.

The latest version of this method can be found at:

http://www.umanitoba.ca/faculties/medicine/human_genetics/gietz/method.html

REFERENCE

Gietz R.D. and Schiestl R.H. 1995. Transforming yeast with DNA. *Methods Mol. Cell. Biol.* **5:** 255–269.

"Lazy Bones" Plasmid Transformation of Yeast Colonies

PROCEDURE

This protocol is modified from Elble (1992).

1. Pick a colony (2–3 mm in diameter) from a plate with a toothpick and transfer cells to a sterile 1.5-ml microfuge tube.

2. Add 10 µl of carrier DNA (100 µg) plus transforming plasmid DNA (in 10 µl) and vortex well. (Carrier DNA does not need to be added if the transforming DNA has come from mini-prep DNA that has not been RNased.)

3. Add 0.5 ml of PLATE solution and vortex.

4. Incubate overnight—4 days at room temperature on the benchtop.

5. Heat shock for 15 minutes at 42°C.

6. Pellet cells for 10 seconds at 8–10 krpm in a microfuge. Carefully remove liquid and gently resuspend cells in 200 µl of sterile distilled H_2O by pipetting up and down. Spread mixture directly onto selective plates.

MATERIALS AND SOLUTIONS

PLATE solution
   40% Polyethylene glycol (PEG) (MW 3350; Sigma P 3640)
   0.1 M Lithium acetate (LiAc)
   10 mM Tris-HCl (pH 7.5)
   1 mM EDTA
Carrier DNA (10 µg/µl)

TIP:

Cells from a fresh plate transform very efficiently; old cells that have been on plates several months can also be transformed, but less efficiently.

REFERENCE

Elble R. 1992. A simple and efficient procedure for transformation of yeasts. *BioTechniques* **13:** 18–20.

Yeast DNA Isolations

A. Yeast DNA Miniprep (40 ml)

PROCEDURE

1. Grow cells at 30°C to saturation (overnight) in 40 ml of YPD in a 125-ml flask.

2. Centrifuge the cells in a clinical centrifuge or a Sorvall SS-34 rotor at 5000 rpm for 5 minutes in a screw-capped centrifuge tube. Discard the supernatant.

3. Resuspend the cells in 3 ml of 0.9 M sorbitol, 0.1 M Na_2 EDTA (pH 7.5).

4. Add 0.1 ml of a 2.5 mg/ml solution of Zymolyase 100T and incubate for 1 hour at 37°C.

5. Centrifuge the cells in a clinical centrifuge or Sorvall SS-34 rotor for 5 minutes at 5000 rpm. Discard the supernatant.

6. Resuspend the cell pellet in 5 ml of 50 mM Tris-Cl (pH 7.4), 20 mM Na_2 EDTA.

7. Add 0.5 ml of 10% SDS and mix.

8. Incubate for 30 minutes at 65°C.

9. Add 1.5 ml of 5 M potassium acetate and store on ice for 1 hour.

10. Centrifuge in a Sorvall SS-34 rotor at 10,000 rpm for 10 minutes.

11. Transfer the supernatant to a fresh plastic centrifuge tube and add two volumes of 95% ethanol at room temperature. Mix and centrifuge at 5000–6000 rpm for 15 minutes at room temperature.

12. Discard the supernatant. Dry the pellet and then resuspend in 3 ml of TE (pH 7.4). This may take several hours.

13. Centrifuge in a Sorvall SS-34 rotor at 10,000 rpm for 15 minutes and transfer the supernatant to a new tube. Discard the pellet.

14. Add 150 µl of a 1 mg/ml solution of RNase A and incubate for 30 minutes at 37°C.

15. Add one volume of 100% isopropanol and shake gently to mix. Remove the precipitate, which should now look like a loose "cocoon" of fibers. Do not centrifuge. Air-dry.

16. Resuspend the precipitate in 0.5 ml of TE (pH 7.4). Store at 4°C. The final concentration of yeast DNA should be ~200 µg/ml. If the final solution is milky, reprecipitate the DNA with isopropanol or centrifuge in a Sorvall SS-34 rotor at 10,000 rpm for 15 minutes.

MATERIALS AND SOLUTIONS

125-ml flask containing 40 ml of YPD

Screw-capped centrifuge tubes

0.9 M Sorbitol, 0.1 M Na$_2$ EDTA (pH 7.5) (3 ml)

Zymolyase 100T (120493-1, Seikagaku America Inc.) solution (0.1 ml)

2.5 mg/ml in 0.9 M sorbitol, 0.1 M Na$_2$ EDTA (pH 7.5)

50 mM Tris-Cl (pH 7.4), 20 mM Na$_2$ EDTA (5 ml)

10% SDS

5 M Potassium acetate (1.5 ml)

95% Ethanol

TE (pH 7.4)

   10 mM Tris-Cl (pH 7.4)

   1 mM Na$_2$ EDTA

RNase A solution (150 μl)

   Dissolve at a concentration of 1 mg/ml in 50 mM potassium acetate (pH 5.5). Boil
   for 10 minutes. Store frozen at –20°C.

100% Isopropanol

B. YEAST DNA MINIPREP (5 ml)

PROCEDURE

1. Grow cells overnight at 30°C in 5 ml of YPD.

2. Collect the cells in a clinical centrifuge at 2000 rpm for 5 minutes. Discard the supernatant.

3. Resuspend the cells in 0.5 ml of 1 M sorbitol, 0.1 M Na$_2$ EDTA (pH 7.5); transfer to a 1.5-ml microfuge tube.

4. Add 0.02 ml of a 2.5 mg/ml solution of Zymolyase 100T and incubate for 1 hour at 37°C.

5. Centrifuge in a microfuge for 1 minute.

6. Discard the supernatant. Resuspend the cells in 0.5 ml of 50 mM Tris-Cl (pH 7.4), 20 mM Na$_2$ EDTA.

7. Add 0.05 ml of 10% SDS; mix well.

8. Incubate the mixture for 30 minutes at 65°C.

9. Add 0.2 ml of 5 M potassium acetate and place the microfuge tube in ice for 1 hour.

10. Centrifuge in microfuge for 5 minutes.

11. Transfer the supernatant to a fresh microfuge tube and add one volume of 100% isopropanol at room temperature. Mix and allow it to sit at room temperature for

5 minutes. Centrifuge VERY BRIEFLY (10 seconds) in a microfuge. Pour off the supernatant and air-dry the pellet.

12. Resuspend the pellet in 0.3 ml of TE (pH 7.4).

13. (Optional) Add 15 μl of a 1 mg/ml solution of RNase A and incubate for 30 minutes at 37°C.

14. Add 0.03 ml of 3 M sodium acetate and mix. Precipitate with 0.2 ml of 100% isopropanol. Centrifuge briefly again to collect the pellet of DNA.

15. Pour off the supernatant and air-dry. Resuspend the pellet of DNA in 0.1–0.3 ml of TE (pH 7.4).

16. Before using the DNA solution in a restriction digest, it may be necessary to centrifuge the final solution hard (15 minutes) in a microfuge to remove insoluble material that may inhibit digestion.

MATERIALS AND SOLUTIONS

YPD

1 M Sorbitol, 0.1 M Na$_2$ EDTA (pH 7.5) (0.5 ml)

Zymolyase 100T (120493-1, Seikagaku America Inc.) solution (0.02 ml)

2.5 mg/ml in 1 M sorbitol, 0.1 MNa$_2$ EDTA (pH 7.5)

50 mM Tris-Cl (pH 7.4), 20 mM Na$_2$ EDTA (0.5 ml)

10% SDS

5 M Potassium acetate (0.2 ml)

100% Isopropanol

TE (pH 7.4)

    10 mM Tris-Cl (pH 7.4)

    1 mM Na$_2$ EDTA

RNase A solution (15 μl) (optional)

Dissolve at a concentration of 1 mg/ml in 50 mM potassium acetate (pH 5.5). Boil for 10 minutes. Store frozen at –20°C.

3 M Sodium acetate (0.03 ml)

C. A Ten-Minute DNA Preparation from Yeast

Modified from Hoffman and Winston (1987).

SAFETY NOTES

Phenol is highly corrosive and can cause severe burns. Gloves, protective clothing, and safety glasses should be worn when handling it. All manipulations should be carried out in a chemical hood. Any areas of skin that come in contact with phenol should be rinsed with a large volume of water or polyethylene glycol 400 and washed with soap and water; ethanol should <u>not</u> be used.

Chloroform is irritating to the skin, eyes, mucous membranes, and upper respiratory tract. It should only be used in a chemical hood. Gloves and safety goggles should also be worn. Chloroform is a carcinogen and may damage the liver and kidneys.

Nitric acid is volatile and should be used in a hood. Concentrated acids should be handled with great care; gloves and a face protector should be worn.

PROCEDURE

Release of Plasmid for Transformation of *E. coli* or Yeast

1. Grow small cultures (at least 1.4 ml) overnight at 30°C in a medium that maintains selection for the plasmid DNA, such as SC–ura.

2. Fill a 1.5-ml microfuge tube with the culture and collect the cells by a 5-second centrifugation in a microfuge.

3. Decant the supernatant and briefly vortex the tube to resuspend the pellet in the residual liquid.

4. Add 0.2 ml of 2% Triton X-100, 1% SDS, 100 mM NaCl, 10 mM Tris-Cl (pH 8), 1 mM Na$_2$ EDTA. Add 0.2 ml of phenol:chloroform:isoamyl alcohol (25:24:1). Add 0.3 g of acid-washed glass beads.

5. Vortex for 2 minutes.

6. Centrifuge for 5 minutes in a microfuge.

7. Transform 0.2 ml of competent *E. coli* cells with 1–5 μl of the aqueous layer. Transform yeast with 15 μl of the aqueous phase.

Isolation of Genomic DNA for Southern Blot Analysis

1. Grow 10-ml yeast cultures to saturation in YPD at 30°C.

2. Collect the cells by centrifugation for 2 minutes in a clinical centrifuge. Remove the supernatant and resuspend the cells in 0.5 ml of distilled H$_2$O. Transfer the cells to a 1.5-ml microfuge tube and collect them by centrifugation for 5 seconds in a microfuge.

3. Follow step 3 above.

4. Follow step 4 above.

5. Vortex for 3–4 minutes. Add 0.2 ml of TE (pH 8).

6. Centrifuge for 5 minutes in a microfuge. Transfer the aqueous layer to a fresh tube. Add 1 ml of 100% ethanol. Mix by inversion.

7. Centrifuge for 2 minutes in a microfuge. Discard the supernatant. Resuspend the pellet in 0.4 ml of TE plus 3 μl of a 10 mg/ml solution of RNase A. Incubate for 5 minutes at 37°C. Add 10 μl of 4 M ammonium acetate plus 1 ml of 100% ethanol. Mix by inversion.

8. Centrifuge for 2 minutes in a microfuge. Discard the supernatant. Air-dry the pellet and resuspend in 50 μl of TE. Use 10 μl for each sample to be analyzed by Southern blotting. This is ~2–4 μg of DNA.

MATERIALS AND SOLUTIONS

Medium that maintains selection for the plasmid DNA, such as SC–ura
2% Triton X-100, 1% SDS, 100 mM NaCl, 10 mM Tris-Cl (pH 8), 1 mM Na$_2$ EDTA (0.4 ml)
Phenol:chloroform:isoamyl alcohol (25:24:1) (0.4 ml)
Glass beads
     0.45–0.5-mm beads are available from a variety of suppliers (e.g., Sigma, American QUALEX, Midwest Scientific, Stratagene). Beads can be cleaned by soaking in nitric acid and washing in copious amounts of distilled H$_2$O. Beads should be dried before use.
YPD
Sterile distilled H$_2$O
TE (pH 8)
     10 mM Tris-Cl (pH 8)
     1 mM Na$_2$ EDTA
100% Ethanol
RNase A stock solution (3 μl)
     Dissolve at a concentration of 10 mg/ml in 50 mM potassium acetate (pH 5.5). Boil for 10 minutes. Store frozen at –20°C.
4 M Ammonium acetate (10 μl)

REFERENCES

Hoffman C.S. and Winston F. 1987. A ten-minute DNA preparation from yeast efficiently releases autonomous plasmids for transformation of *Escherichia coli. Gene* **57**: 267–272.

D. Yeast Genomic DNA: Glass Bead Preparation

1. Grow cells overnight in 5 ml of medium (rich or selective) at 30°C in a roller drum.

2. Transfer cultures to 13 x 100-mm glass tubes, and spin down cells in a table-top centrifuge at 1500 rpm for 5 minutes.

3. Wash cells with 3 ml of sterile H$_2$O, and spin as above.

4. Resuspend in 500 μl of Lysis buffer.

5. Add clean glass beads (about two-thirds of a 1.5-ml Eppendorf tube) and 25 μl of 5 M NaCl.

6. Vortex on highest setting for 1 minute.

7. Spin at 2 krpm for 2 minutes.

8. Transfer the liquid with a P-1000 to a 1.5-ml Eppendorf tube.

9. Add 500 μl of phenol, vortex, and spin for 1 minute. Extract aqueous layer (top) with a P-1000 and transfer to a clean tube. Add 500 μl of SEVAG (24:1 chloroform:isoamyl alcohol), vortex, spin, and extract as above.

10. Add 1 ml of cold 95% ethanol and precipitate for 1 hour at –20°C.

11. Pellet the DNA by spinning for 5 minutes at full speed, pour off the supernatant, and wash with 70% ethanol. Resuspend in 250 μl of TE.

12. Add 25 μl of EDTA-Sark and 5 μl of proteinase K (10 mg/ml). Incubate for 30 minutes at 37°C.

13. Add 250 μl of 5 M NH$_4$Ac, and repeat steps 9 and 10.

14. Pellet the DNA, wash with 70% ethanol, and resuspend in 100 μl of TE (use ~10 μl/digest).

MATERIALS AND SOLUTIONS

YPD
13 x 100-mm glass tubes
Sterile H$_2$O
Lysis buffer
    0.1 M Tris-Cl (pH 8.0)
    50 mM EDTA
    1% SDS
5 M NaCl
Glass beads (0.5 mm in diameter; BioSpec Products 11079-105)
SEVAG (24:1 chloroform:isoamyl alcohol)
95% Ethanol
70% Ethanol
EDTA-Sark
    0.4 M EDTA (pH 8.0)
    2% N-lauroylsarcosine (Sarkosyl)
Proteinase K (10 mg/ml)
5 M NH$_4$(C$_2$H$_3$O$_2$)
TE
    10 mM Tris-Cl
    1 mM EDTA (pH 8.0)
Phenol, equilibrated with H$_2$O (see Safety Notes in part C)

Yeast Protein Extracts

PROCEDURE

This procedure is useful for making yeast protein extracts for SDS gel electrophoresis and Western blotting.

1. The extract is made from 2 OD_{600} units of cells. An easy way to prepare a culture is to inoculate 5 ml of YPD with a very small mass of cells from a plate. This should give an exponential culture (OD_{600} = 0.5–2.0) after overnight growth.

2. Transfer 2 OD_{600} units of cell culture to a 13-mm tube containing 2 ml of 50 mM Tris (pH 7.5), 10 mM NaN_3 on ice. Pellet cells in a swinging-bucket clinical centrifuge.

3. Aspirate supernatant with a 25-gauge needle and suspend cells in 30 μl of ESB.

4. Quickly transfer to a microcentrifuge tube and heat to 100°C for 3 minutes. This heating step should be done rapidly to inactivate proteases. The samples can be stored at –20°C after this.

5. Add about 0.1 g of 0.2-mm glass beads, or pour beads until they reach the top of the liquid. Agitate by vigorous vortex mixing for 2 minutes.

6. Add 70 μl of ESB, vortex briefly, and heat to 100°C for 1 minute.

7. Load 5–20 μl of the extract on an SDS gel.

MATERIALS AND SOLUTIONS

Appropriate growth medium
50 mM Tris (pH 7.5), 10 mM NaN_3
ESB (can be stored at 4°C for months)
2% SDS
80 mM Tris (pH 6.8)
10% Glycerol
1.5% DTT
0.1 mg/ml Bromophenol blue

TIP:

Quick boiling of the sample is usually sufficient to stabilize proteins against proteolysis. If proteolysis is suspected, PMSF can be added to ESB to reduce proteolysis. In cases of extremely labile proteins, all of the following inhibitors have been found to be beneficial:

Inhibitor	Final concentration	Stock
PMSF	1 mM	100x in isopropanol
pepstatin A	0.7 µg/ml	2000x in methanol
leupeptin	0.5 µg/ml	1000x in H_2O
E64	10 µg/ml	1000x in 50% ethanol
aprotinin	2 µg/ml	5000x in H_2O
α_2 macroglobulin	0.5 U/ml	added directly

For very hydrophobic proteins such as multispanning integral membrane proteins, boiling of the sample will cause aggregation. For these proteins, use a protease inhibitor cocktail in ESB and warm the sample to 37°C instead of boiling.

Yeast RNA Isolation

This procedure is designed to yield more than 10 mg of total nucleic acid and 300–400 µg of poly(A)-selected mRNA from each 500-ml culture. The volumes indicated can be applied directly to cultures of 200–500 ml, but should be adjusted for cultures of significantly smaller or larger volumes.

It is convenient to inoculate 200 ml of YPD (or SC) with an appropriate volume of a fresh overnight culture to yield a concentration of 2×10^7–4×10^7 cells/ml the following morning. The cell density should be checked using a hemocytometer (see Appendix G) or Klett (50–100 Klett units for most strains).

PROCEDURE

SAFETY NOTES

Nitric acid is volatile and should be used in a hood. Concentrated acids should be handled with great care; gloves and a face protector should be worn.

Phenol is highly corrosive and can cause severe burns. Gloves, protective clothing, and safety glasses should be worn when handling it. All manipulations should be carried out in a chemical hood. Any areas of skin that come in contact with phenol should be rinsed with a large volume of water or polyethylene glycol 400 and washed with soap and water; ethanol should <u>not</u> be used.

Diethyl pyrocarbonate is toxic and volatile. Work with open tubes in a hood and wear gloves.

Chloroform is irritating to the skin, eyes, mucous membranes, and upper respiratory tract. It should only be used in a chemical hood. Gloves and safety goggles should also be worn. Chloroform is a carcinogen and may damage the liver and kidneys.

Isolation of Total Yeast RNA

1. Add 50 µg of cycloheximide for each ml of culture. Shake for 15 minutes at 30°C. This step is optional but is supposed to protect mRNA by freezing it in polysomes.

2. To quickly cool the cells for harvesting, fill large centrifuge bottles (Sorvall GS3 or equivalent) halfway with crushed ice, pour the culture over the ice, and shake. Centrifuge at 5000 rpm for 5 minutes at 4°C.

3. Add 11 g of acid-washed glass beads to a 30-ml centrifuge tube. Corex glass tubes work best, but plastic tubes can be used.

4. Add 3 ml of phenol equilibrated with LETS buffer to the glass beads.

5. Resuspend the cell pellet in 2.5 ml of ice-cold LETS buffer and add it to the glass beads/phenol. The meniscus should be just above the surface of the glass beads for the best cell breakage.

6. Vortex at top speed, alternating 30 seconds of vortexing with 30 seconds on ice, for a total of 3 minutes of vortexing. Cell breakage can be checked with a phase-contrast microscope. Broken cells appear as nonrefractile "ghosts."

7. When at least 90% of the cells are broken, add 5 ml of ice-cold LETS buffer and vortex briefly. Centrifuge for 5 minutes at 8000 rpm in a Sorvall SS-34 rotor to break the phases. After centrifugation, the phenol layer plus interface should be just below the surface of the beads, allowing easy retrieval of the aqueous phase without disturbing the interface.

8. Transfer the aqueous phase to a clean tube and extract twice with 5 ml of phenol:chloroform:isoamyl alcohol (25:24:1). Avoid transferring the interface. Extract once with chloroform (optional).

9. Add 1/10 volume of 5 M LiCl and precipitate for 3 hours at –20°C. The RNA precipitate may be conveniently stored at –20°C at this point.

Isolation of Poly(A)+ RNA

1. Centrifuge at 10,000 rpm for 10 minutes in a Sorvall SS-34 rotor to pellet the stored RNA. Wash with 80% ethanol. Dry under a vacuum and dissolve in 2.5 ml of distilled H_2O. When dissolved, add 2.5 ml of 2x loading buffer and add SDS to a final concentration of 0.3%.

2. Heat for 5 minutes at 65°C.

3. Load an oligo(dT) column and wash three times with 1x loading buffer.

4. Elute the poly(A)+ RNA with 10 mM HEPES (pH 7); add RNase inhibitor if desired.

5. Determine the RNA concentration by measuring the OD_{260} of a 1:10 dilution of the poly(A)+ RNA in distilled H_2O. Compare this with a 1:10 dilution of 10 mM HEPES (pH 7) in H_2O (i.e., compare with 1 mM HEPES).

6. Precipitate the RNA by adding 1/10 volume of 3 M sodium acetate and 2.5 volumes of 100% ethanol. Mix by inversion and incubate for 30 minutes at –70°C. Centrifuge for 10 minutes in a microfuge. Remove the supernatant and dissolve the RNA pellet in distilled H_2O to make a final concentration of 1 mg/ml. Both the dissolved RNA or the ethanol precipitate can be stored indefinitely at –70°C.

MATERIALS

Cycloheximide (50 µg/ml of culture) (optional)

Large centrifuge bottles (Sorvall GS3 or equivalent)

Glass beads

    0.45–0.5-mm beads are available from a variety of suppliers (e.g., Sigma, American QUALEX, Midwest Scientific, Stratagene). Beads can be cleaned by soaking in nitric acid and washing in copious amounts of distilled H_2O. Beads should be dried before use.

Centrifuge tubes (Corex glass or plastic; 30-ml)

Phenol equilibrated with LETS buffer (3 ml)

LETS buffer

    0.1 M Lithium chloride (LiCl)

    0.01 M Na_2 EDTA

    0.01 M Tris-Cl (pH 7.4)

    0.2% SDS

    0.1% Diethyl pyrocarbonate (optional)

Phenol:chloroform:isoamyl alcohol (25:24:1)

Chloroform (optional)

5 M LiCl

80% Ethanol

Sterile distilled H_2O

1x Loading buffer

    0.5 M NaCl

    0.01 M HEPES (pH 7)

SDS

Oligo(dT) column

10 mM HEPES (pH 7)

RNase inhibitor (optional)

3 M Sodium acetate

100% Ethanol

Hydroxylamine Mutagenesis of Plasmid DNA

SAFETY NOTES

Hydroxylamine is a mutagen and should be handled carefully. Gloves should be worn when handling it.

Solid NaOH is caustic and should be handled with great care; gloves and face protector should be worn.

1. Prepare the hydroxylamine solution just before use and store on ice until needed.

2. Add 10 µg of CsCl-purified plasmid DNA to 500 µl of hydroxylamine solution in a microfuge tube.

3. Incubate for 20 hours at 37°C.

4. Stop the reaction by adding 10 µl of 5 M NaCl, 50 µl of 1 mg/ml BSA, and 1 ml of 100% ethanol; precipitate the DNA for 10 minutes at –70°C.

5. Centrifuge the precipitated DNA in a microfuge for 10 minutes. Carefully remove all of the supernatant.

6. Resuspend the DNA in 100 µl of TE (pH 8). Add 10 µl of 3 M sodium acetate and 250 µl of 100% ethanol; precipitate the DNA for 10 minutes at –70°C and centrifuge as in step 5.

7. Allow the pellet to air-dry and then resuspend it in 100 µl of TE (pH 8).

8. The DNA can be used directly for transformation of either *E. coli* or yeast. The transformation frequency in yeast with the mutagenized DNA is reduced only approximately threefold relative to unmutagenized DNA. The formation of Ura⁻ plasmids in *E. coli* can be monitored. The *E. coli* strain can be *ung⁺*, but the transformation frequency will be reduced 10- to 100-fold. When the *URA3* gene makes up approximately 10% of the plasmid DNA and an *E. coli ung⁺ pyrF⁻* strain (DB6507) is being transformed, approximately 4% Ura⁻ colonies can be expected. Transforming yeast with the same stock of mutagenized DNA gives approximately 1% loss-of-function mutations ("knockouts") in my favorite gene. Approximately 10% of the mutants are temperature-sensitive.

MATERIALS AND SOLUTIONS

Hydroxylamine solution
   0.35 g of Hydroxylamine HCl
   0.09 g of NaOH
   5 ml of Distilled H_2O (ice-cold)
   Dissolve the solids in the H_2O. The pH should be ~7. Prepare just before use and store on ice until needed.
CsCl-purified plasmid DNA (10 μg)
5 M NaCl
1 mg/ml Bovine serum albumin (BSA)
100% Ethanol
TE (pH 8)
   10 mM Tris-Cl (pH 8)
   1 mM Na_2 EDTA
3 M Sodium acetate

Assay of β-Galactosidase in Yeast

There are two basic methods for the in vitro assay of β-galactosidase from yeast. They differ mainly in the method of preparing the material for assay. In the first method (Rose and Botstein 1983), a crude extract is prepared, and the activity is normalized to the amount of protein assayed. In the second method (Guarente 1983), the cells are permeabilized to allow the substrate to enter the cells, and the activity is normalized to the number of cells assayed. The former method is preferable when comparing cells that are grown under very different conditions or that have different genetic backgrounds. The latter method is adapted from the assay for *E. coli* and is particularly suited for changing levels of activity within a single strain.

METHOD 1: ASSAY OF CRUDE EXTRACTS

SAFETY NOTES

Nitric acid is volatile and should be used in a hood. Concentrated acids should be handled with great care; gloves and a face protector should be worn.

Phenylmethylsulfonyl fluoride (PMSF) is extremely destructive to the mucous membranes of the respiratory tract, the eyes, and the skin. It may be fatal if inhaled, swallowed, or absorbed through the skin. In case of contact, immediately flush eyes or skin with copious amounts of water and discard contaminated clothing.

Chloroform is irritating to the skin, eyes, mucous membranes, and upper respiratory tract. It should only be used in a chemical hood. Gloves and safety goggles should also be worn. Chloroform is a carcinogen and may damage the liver and kidneys.

1. Grow a 5-ml culture of cells to a concentration of 1×10^7–2×10^7 cells/ml in an appropriate liquid medium at an appropriate temperature (usually 30°C). If the hybrid gene is expressed from an autonomous plasmid, use an appropriate medium to select for the presence of the plasmid.

2. Chill the cells on ice and harvest by centrifugation (2000 rpm for 5 minutes in a clinical centrifuge is adequate).

Keep the cells on ice from this point on.

3. Discard the supernatant. Resuspend the cells in 250 µl of breaking buffer. The cells can now be frozen at –20°C and assayed at a later date.

All of the following steps can be performed in a 1.5-ml microfuge tube.

4. If the cells were frozen, thaw them on ice. Add glass beads until the beads reach a level just below the meniscus of the liquid. Add 12.5 µl of PMSF stock solution.

5. Vortex six times at top speed in 15-second bursts. Chill on ice between bursts.

6. Add 250 µl of breaking buffer and mix well. Withdraw the liquid extract after plunging the tip of a 1000-µl pipettor to the bottom of the tube.

7. Clarify the extract by centrifuging for 15 minutes in a microfuge. If the activity is in the particulate fraction, the unclarified supernatant can be used and the assay mixture may be clarified later in step 8.

8. To perform the assay:

   a. Add 10–100 µl of extract to 0.9 ml of Z buffer. Adjust the volume to 1 ml with breaking buffer.

   b. Incubate the mixture in a water bath at 28°C for 5 minutes.

   c. Initiate the reaction by adding 0.2 ml of ONPG stock solution. Note precisely the time that the addition is made. Incubate at 28°C until the mixture has acquired a pale yellow color.

   d. Terminate the reaction by adding 0.5 ml of Na_2CO_3 stock solution. Note precisely the time that the reaction is terminated. Measure the optical density at 420 nm.

9. Measure the protein concentration in the extract using the dye-binding assay of Bradford (1976):

   a. Dilute the Bradford reagent fivefold in distilled H_2O. Filter the diluted reagent through Whatman 540 paper (or equivalent).

   b. Add 10–20 µl of the extract to 1 ml of the diluted reagent and mix. Measure the blue color formed at 595 nm. Use disposable plastic cuvettes to prevent the formation of a blue film.

   c. Prepare a standard curve using several dilutions (0.1–1 mg/ml) of BSA dissolved in breaking buffer.

   Typical extracts prepared in this fashion contain 0.5–1 mg/ml of protein.

10. Express the specific activity of the extract according to the following formula:

$$\frac{OD_{420} \times 1.7}{0.0045 \times \text{protein} \times \text{extract volume} \times \text{time}}$$

OD_{420} is the optical density of the product, *o*-nitrophenol, at 420 nm. The factor 1.7 corrects for the reaction volume. The factor 0.0045 is the optical density of a 1 nmole/ml solution of *o*-nitrophenol. Protein concentration is expressed as mg/ml. Extract volume is the volume assayed in ml. Time is in minutes. Specific activity is expressed as nmoles/minute/mg protein.

METHOD II: PERMEABILIZED CELL ASSAY

1. Grow the cells as above. Measure the OD_{600} of the culture and harvest 1×10^6–1×10^7 cells by centrifugation as above.

2. Discard the supernatant. Resuspend the cells in 1 ml of Z buffer.

3. Add 3 drops of chloroform and 2 drops of 0.1% SDS. Vortex at top speed for 10 seconds.

4. Preincubate the samples for 5 minutes at 28°C. Start the reaction by adding 0.2 ml of ONPG as above.

5. Stop the reaction by adding 0.5 ml of Na_2CO_3 stock solution when the sample in the tube has developed a pale yellow color. Note the amount of time elapsed during the assay. Remove the cell debris by centrifuging for 10 minutes in a microfuge and then discarding the pellet.

6. Measure the OD_{420} of the reactions.

7. Express the activity as β-galactosidase units:

$$\frac{OD_{420}}{OD_{600} \text{ of assayed culture} \times \text{volume assayed} \times \text{time}}$$

OD_{420} is the optical density of the product, o-nitrophenol. OD_{600} is the optical density of the culture at the time of assay. Volume is the amount of the culture used in the assay in ml. Time is in minutes.

MATERIALS AND SOLUTIONS

Appropriate liquid medium

Breaking buffer

    100 mM Tris-Cl (pH 8)

    1 mM Dithiothreitol

    20% Glycerol

Glass beads

    0.45–0.5-mm beads are available from a variety of suppliers (e.g., Sigma, American QUALEX, Midwest Scientific, Stratagene). Beads can be cleaned by soaking them in nitric acid and then washing them in copious amounts of distilled H_2O. Beads should be dried before use.

PMSF (Sigma P 7626) stock solution

    40 mM in 100% isopropanol. Store at –20°C.

Z buffer (Miller 1972)

    16.1 g of $Na_2HPO_4 \cdot 7H_2O$

    5.5 g of $NaH_2PO_4 \cdot H_2O$

    0.75 g of KCl

0.246 g of $MgSO_4 \cdot 7H_2O$

2.7 ml of β-Mercaptoethanol

Distilled H_2O to make a final volume of 1 liter

Adjust the pH to 7. Store at 4°C.

ONPG (*o*-nitrophenyl-β-D-galactoside) stock solution

4 mg/ml in Z buffer. Store at –20°C.

Na_2CO_3 stock solution

1 M in distilled H_2O

Bradford reagent (Bio-Rad)

Distilled H_2O

Whatman 540 paper or equivalent

Disposable plastic cuvettes

0.1–1 mg/ml Bovine serum albumin (BSA) in breaking buffer

Chloroform

0.1% SDS

REFERENCES

Bradford M.M. 1976. A rapid and sensitive method for the quantitation of microgram quantities of protein utilizing the principle of protein-dye binding. *Anal. Biochem.* **72:** 248–254.

Guarente L. 1983. Yeast promoters and *lacZ* fusions designed to study expression of cloned genes in yeast. *Methods Enzymol.* **101:** 181–191.

Miller J.H. 1972. *Experiments in molecular genetics.* Cold Spring Harbor Laboratory, Cold Spring Harbor, New York.

Rose M. and Botstein D. 1983. Construction and use of gene fusions *lacZ* (β-galactosidase) that are expressed in yeast. *Methods Enzymol.* **101:** 167–180.

Plate Assay for Carboxypeptidase Y

PROCEDURE

This protocol was adapted from Jones (1991).

1. Grow colonies or patches on YPD plates (3 days for colonies and 1 day for patches is usually sufficient).

2. Carefully pour overlay solution over the surface of the plate to completely cover the cells.

3. After the agar hardens in 5–10 minutes, carefully flood the surface with *fresh* Fast Garnet GBC solution.

4. Allow color to develop for 5 minutes. Wild-type strains will turn red, and carboxypeptidase Y-negative strains will appear yellow or pink.

5. Decant Fast Garnet solution to best observe developed color.

MATERIALS AND SOLUTIONS

Overlay solution for one plate

In glass or polypropylene tube, mix 2.5 ml of 1 mg/ml N-acetyl-DL-phenylalanine β-naphthyl ester in dimethylformamide with 4 ml of 0.6% molten agar. Hold at 50°C.

Fast Garnet GBC solution (use 5 ml/plate)

5 mg/ml Fast Garnet GBC (sulfate salt; Sigma F 8761) in 0.1 M Tris-HCl (pH 7.4)

Dimethylformamide permeabilizes the cells. Carboxypeptidase Y in the cells cleaves the ester linkage in N-acetyl-DL-phenylalanine β-naphthyl ester. Free β-naphthol then reacts with the diazonium salt Fast Garnet GBC to produce an insoluble red dye.

REFERENCE

Jones E.W. 1991. Tackling the protease problem in *Saccharomyces cerevisiae*. *Methods Enzymol.* **194**: 428–453.

Random Spore Analysis

PROCEDURE

I. Sporulation

1. Patch out a single colony of the diploid to be sporulated onto YPD. It is best to spread the cells as thinly as possible. Incubate for 12–16 hours at 30°C. Growth for more than 16 hours will dramatically decrease the efficiency of sporulation.

2. In the morning, with a sterile dowel transfer a matchhead quantity of cells to a test tube containing 2.5 ml of sporulation medium and place on a rotor at 25°C.

3. Monitor the extent of sporulation by light microscopy. It can take from 2–10 days for more than 5% of cells to sporulate.

II. Random Spores

4. Transfer 1 ml of sporulated culture to a 15-ml conical polystyrene tube and collect the cells by centrifugation for 5 minutes in the clinical centrifuge.

5. Remove the supernatant completely and resuspend cells in 0.2 ml of sterile H_2O and 5 µl of β-glucuronidase (~500 units).

6. Incubate on a rotor for 1 hour at 30°C.

7. Add 0.1 ml (~0.15 g) of sterile 0.5-mm glass beads. Incubate on a rotor for 1 hour at 30°C.

8. Add 1 ml of sterile H_2O.

9. Vortex 1–2 minutes and check by light microscopy for complete disruption of asci.

10. Add 4 ml of sterile H_2O.

11. Make dilutions of 10^{-1} to 10^{-3} in sterile H_2O. Plate 200 µl onto SC-arg with 60 µg/ml canavanine.

MATERIALS AND SOLUTIONS

Sporulation medium
     1% KOAc
     0.025% Glucose
β-Glucuronidase (Sigma G 7770)
Sterile 0.5-mm glass beads
SC-arg with 60 μg/ml canavanine plates

Yeast Vital Stains

PROCEDURE

Nuclear and Mitochondrial DNA

SAFETY NOTES

DAPI is a possible carcinogen. It may be harmful if inhaled, swallowed, or absorbed through the skin. It may also cause irritation. Wear gloves, face mask, and safety glasses and do not breathe the dust.

1. Pellet ~10^7 cells in a microfuge tube (5 second pulse) and resuspend in 70% ethanol.

2. Fix for 5 minutes or more and wash twice with H_2O.

3. Suspend cells in a small volume of 50 ng/ml DAPI (4′,6-diamidino-2-phenylindole; Sigma D 9542 or Accurate Chemical and Scientific Corp.) in mounting medium. A stock of 1 mg/ml in H_2O can be stored at –20°C.

4. Observe with UV filter set.

Cells fixed in formaldehyde can also be stained with DAPI in mounting medium. DAPI can also be used as a vital stain of cells in growth medium at a concentration of ~1 µg/ml but the background staining of the cell bodies is higher than with fixed cells.

Mitochondria

1. To ~10^7 cells in growth medium, add 100 ng/ml $DiOC_6$ (3,3′-dihexyloxacarbocyanine iodide; Sigma D 3652 or Molecular Probes) (dilute 1/10^4 from 1 mg/ml stock in ethanol, which is stable for months in the dark at –20°C).

2. Incubate for 5 minutes or more and observe with fluorescein filter set.

The concentration of $DiOC_6$ may need to be optimized—at ~1 µg/ml all membranes appear to be stained.

Vacuole

1. Pellet ~10^7 cells in a microfuge tube (5-second pulse) and resuspend in YPD with 50 mM sodium citrate (pH 4.0).

2. Add CDCFDA (carboxy-2´,7´-dichloro-fluorescein diacetate; Molecular Probes) to 10 μM (1/1000 dilution of a 10 mM stock in dimethylformamide, which is stable for months at –20°C).

3. Incubate for 10 minutes or more and observe with fluorescein filter set.

Bud Scars and Chitin

1. To ~10^7 cells in growth medium, add 100 μg/ml Calcofluor (Fluorescent brightener 28; Sigma F 3397) (dilute 1/10 from 1 mg/ml stock in H$_2$O, which is stable for weeks in the dark at –20°C).

2. Incubate 5 minutes or more, wash twice with H$_2$O, and observe with UV filter set.

Yeast Immunofluorescence

PROCEDURE

Preparation of Cells

SAFETY NOTES

Formaldehyde is toxic and is a carcinogen. It is readily absorbed through the skin and is irritating to the eyes, skin, mucous membranes, and upper respiratory tract. Wear gloves and safety glasses and always work in a chemical hood.

1. Grow a 5-ml culture to early exponential phase (10^6–10^7 cells/ml).

2. Fix cells by adding 1/10 volume formaldehyde directly to medium (total formaldehyde concentration 3.7%; standard stock solution is 37%).

3. Incubate cells in formaldehyde for at least 1 hour.

4. Pellet cells and wash once with 0.1 M potassium phosphate (pH 7.5).

5. Resuspend cells in 1 ml of 50 μg/ml Zymolyase 100T or ~50 units/ml lyticase in 0.1 M potassium phosphate (pH 7.5) with 2 μl/ml β-mercaptoethanol.

6. Incubate for 30 minutes at 30°C. Check efficiency of spheroplasting by phase contrast microscopy. The cells should be a dark, translucent gray. Bright (refractile) cells are insufficiently digested. Ghosts (pale gray with little if any internal structure) have been overdigested.

7. Pellet gently (low speed in microfuge) and resuspend in 1 ml of PBS.

Staining

SAFETY NOTES

DAPI is a possible carcinogen. It may be harmful if it is inhaled, swallowed, or absorbed through the skin. It may also cause irritation. Wear gloves, face mask, and safety glasses, and do not breathe the dust.

1. Prepare Teflon-masked slides (10-well multiwell slides) by putting 10 μl of 1 mg/ml polylysine (size 400,000) onto each well. Wash slide with distilled H_2O and let dry.

2. Place 10 μl of fixed cells onto each well. After a few minutes, aspirate and wash three times with PBS. Check the slide in the microscope to ensure the cells are at a suitable density and not clumped.

3. Optional step (not necessary for the antibodies used in this course, but recommended):

   Dunk the slide in cold methanol (–20°C) for 6 minutes and then into cold acetone (–20°C) for 30 seconds. This treatment results in flatter cells, which can aid visualization of the cytoskeleton. For some antibody-antigen combinations, this step might be necessary for reactivity. Rehydrate wells with PBS.

4. Place 15 μl of blocking buffer consisting of PBS + 3% BSA (bovine serum albumin) on wells. Incubate for 30 minutes in a humid chamber such as a Petri dish with a wet Kimwipe in it. It is important to not let slide wells dry out from this point on. (The BSA reduces nonspecific antibody binding to the slide by blocking protein-binding sites).

5. Remove the blocking buffer by aspiration.

6. Add 10–15 μl of primary antibody in PBS + 3% BSA to the wells and incubate for 1 hour in a humid chamber. Use affinity-purified antibodies or monoclonals, and if possible, use an isogenic control that lacks the antigen of interest. It is also helpful to do a control without primary antibody.

7. Aspirate off primary antibody and wash three times with PBS + 3% BSA.

8. Repeat step 6 with fluorescent secondary antibody (incubate in the dark).

9. Aspirate off secondary antibody and wash three times with PBS + 3% BSA.

10. Add 10–15 μl of 1 μg/ml DAPI to wells, incubate for about 1 minute, and wash three times with PBS.

11. Remove the PBS, and add a small drop of mounting medium to each well. Put on the coverslip, trying to avoid trapping air bubbles in the wells. Remove excess mounting solution with a Kimwipe, taking care not to move the coverslip. Seal with clear nail polish and store at –20°C.

MATERIALS AND SOLUTIONS

Zymolyase 100T (120493-1, Seikagaku America)

Mounting medium

   Dissolve 50 mg of *p*-phenylenediamine in 5 ml of PBS (see below) and adjust to pH 9. Add 45 ml of glycerol and stir until homogeneous. Store in aliquots at –70°C in the dark. (Modified from Johnson and Nogueira Araujo 1981.)

   If DNA-staining is also required, 4´,6-diamidino-2-phenylindole (DAPI; Sigma D 9542 or Accurate Chemical and Scientific Corp.) is included in the mounting medium (Williamson and Fennell 1975). To 50 ml of mounting medium, add 2.5 μl of fresh DAPI solution (1 mg of DAPI/ml H_2O).

PBS

A 20x stock contains per liter:

160 g of NaCl

4 g of KCl

22.8 g of Na_2HPO_4

4 g of KH_2PO_4

Adjust to pH 7.3 with 10 N NaOH.

REFERENCES

Johnson G.D. and Nogueira Araujo G.M. 1981. A simple method of reducing the fading of immuno-fluorescence during microscopy. *J. Immunol. Methods* **43**: 349–350.

Williamson D.H. and Fennell D.J. 1975. The use of fluorescent DNA-binding agent for detecting and separating yeast mitochondrial DNA. *Methods Cell Biol.* **12**: 335–351.

Actin Staining in Fixed Cells

Phalloidin binds specifically to F-actin, and fluorescent-tagged phalloidin stains the actin skeleton in cells in a manner that is very close to the staining pattern seen using anti-actin antibody.

PROCEDURE

SAFETY NOTES

Formaldehyde is toxic and is a carcinogen. It is readily absorbed through the skin and is irritating to the eyes, skin, mucous membranes, and upper respiratory tract. Wear gloves and safety glasses and always work in a chemical hood.

1. Grow cells to exponential phase ($\sim 10^7$ cells/ml).

2. Fix cells in a microfuge tube by adding 0.1 ml of formaldehyde directly to 1 ml of culture medium (formaldehyde concentration 3.7%; standard stock solution is 37%).

3. Incubate cells in formaldehyde for 30 minutes or more.

4. Wash two times with PBS; pelleting with 5-second pulses in microfuge.

5. Suspend cells in 50 µl of PBS and add 5 U of rhodamine or fluorescein-conjugated phalloidin (normally this would be 25 µl of a 200 U/ml stock in methanol, which is stable for months at –20°C).

6. Wash three times with PBS and resuspend in a small volume of mounting medium.

7. Observe with either rhodamine or fluorescein filter sets.

PCR Protocol for PCR-mediated Gene Disruption

PRODCEDURE

This protocol was adapted from Brachmann et al. (1997).

1. Reaction mix:

    5 µl of 10x *Taq* Buffer

    5 µl of 25 mM $MgCl_2$

    2 µl of 10 mM dNTPs

    10–100 ng of template DNA

    25 pmole of each primer

    0.5 µl of *Taq* polymerase (2 U)

    ==> H_2O to 50 µl total volume

2. PCR cycle profile:

    94°C   5 minutes

    94°C   1 minute

    55°C   1 minute

    72°C   2 minutes

    ==> 10 cycles

    94°C   1 minute

    65°C   1 minute

    72°C   2 minutes

    ==> 20 cycles

    72°C   10 minutes

NOTE:

Typically, the entire reaction can be used for a transformation, although purifying the amplified DNA product increases transformation efficiency dramatically. We have success using Wizard PCR Preps DNA Purification System (Promega Corp. A 7170).

MATERIALS AND SOLUTIONS

10x *Taq* Buffer

    0.5 M KCl

    100 mM Tris-Cl (pH 8.5)

    0.1% Triton X-100

25 mM $MgCl_2$

10 mM dNTPs

pRS40X template DNA (mini-prep DNA works well)

Taq polymerase

Two gene-specific DNA primers

    One oligonucleotide should consist of 40 nucleotides of gene-specific sequence for one end of the targeted region at the 5′ end followed by:

        5′-CTGTGCGGTATTTCACACCG-3′ (left primer),

    and another 40 nucleotides homologous to the other side of the targeted region at the 5′ end followed by:

        5′-AGATTGTACTGAGAGTGCAC-3′ (right primer).

    The primers are then used to amplify any auxotrophic marker from a pRS40X or pRS30X integrating plasmid (Sikorski and Hieter 1989; Brachmann et al. 1997).

REFERENCES

Brachmann C.B., Davies A., Cost G.J., Caputo E., Li J., Hieter P., and Boeke J.D. 1997. Designer deletion strains derived from *Saccharomyces cerevisiae* S288C: A useful set of strains and plasmids for PCR-mediated gene disruption and other applications. *Yeast* **14**: 115–132.

Sikorski R.S. and Hieter P. 1989. A system of shuttle vectors and yeast host strains designed for efficient manipulation of DNA in *Saccharomyces cerevisiae. Genetics* **122**: 19–27.

Yeast Colony PCR Protocol

PROCEDURE

1. Combine reaction mix on ice:

    2 µl of 10x Colony PCR Buffer

    1.2 µl of 25 mM $MgCl_2$

    0.4 µl of 10 mM dNTPs

    10 pmole of each primer

    0.2 µl *Taq* polymerase (5 U)

    ==> H_2O to 20 µl

2. Using a pipette tip, add a very small amount of cells (~0.25 µl) into PCR reaction mix (20 µl of reaction).

3. PCR cycle profile:

    94°C   4 minutes

    94°C   1 minute

    55°C   1 minute

    72°C   2 minutes

    ==> 35 cycles

    72°C   10 minutes

4. Load entire sample on agarose gel.

MATERIALS AND SOLUTIONS

10x Colony PCR Buffer

    0.125 M Tris-HCl (pH 8.5)

    0.56 M KCl

25 mM $MgCl_2$

10 mM dNTPs

Taq polymerase

Two gene-specific DNA primers

Each oligonucleotide should be 25 nucleotides long and specific for either side of the region of interest to be amplified.

Note: The elongation times (at 72°C) work well for amplification of loci ≤1.5 kbp in size. These conditions may need to be modified for amplification of longer regions.

Measuring Yeast Cell Density by Spectrophotometry

PROCEDURE

1. Take 1-ml samples of cells from the liquid medium in which they are growing, and place them in microfuge tubes. If necessary, dilute the cells to get an OD_{660} of less than 1.

2. Sonicate the cells to ensure an accurate reading of cell number, then transfer the samples to cuvettes. Prepare a blank using the growth medium in which the cells were grown. Use this sample to zero the absorbance before measuring the optical densities of the samples.

3. Measure the OD_{660} of each sample.

4. Calculate the cell density of haploid cells using the Table 1 below (kindly provided by Lee Hartwell). Cell densities for diploids are half of those for haploids.

NOTE:

There can be some variability among strains. For example, mutants that are abnormally large, like some *cdc* mutants, will scatter more light than wild-type cells at the same cell density. The table below refers to wild-type cells from strain A364A. There can also be some variability between spectrophotometers. It is advisable to carefully determine the cell density of your strain using a Coulter counter or a hemocytometer, and then measure the OD_{660} to construct a standard curve for your strain.

Table 1. *Haploid yeast (A364A) cell density by OD$_{660}$*

OD 660	#haploid cells x 10^7	OD 660	#haploid cells x 10^7	OD 660	#haploid cells x 10^7	OD 660	#haploid cells x 10^7
.000	.000	.500	.700	1.000	1.850	1.500	4.480
.010	.015	.510	.717	1.010	1.890	1.510	4.550
.020	.025	.520	.733	1.020	1.926	1.520	4.625
.030	.040	.530	.750	1.030	1.963	1.530	4.700
.040	.053	.540	.766	1.040	2.000	1.540	4.775
.050	.065	.550	.783	1.050	2.040	1.550	4.850
.060	.078	.560	.800	1.060	2.080	1.560	4.925
.070	.090	.570	.817	1.070	2.120	1.570	5.000
.080	.103	.580	.833	1.080	2.163	1.580	5.075
.090	.115	.590	.850	1.090	2.206	1.590	5.150
.100	.128	.600	.866	1.100	2.250	1.600	5.225
.110	.140	.610	.883	1.110	2.296	1.610	5.300
.120	.153	.620	.900	1.120	2.343	1.620	5.380
.130	.165	.630	.917	1.130	2.390	1.630	5.460
.140	.178	.640	.933	1.140	2.433	1.640	5.540
.150	.190	.650	.950	1.150	2.476	1.650	5.630
.160	.204	.660	.966	1.160	2.520	1.660	5.700
.170	.216	.670	.983	1.170	2.566	1.670	5.800
.180	.229	.680	1.000	1.180	2.613	1.680	5.890
.190	.241	.690	1.023	1.190	2.660	1.690	5.980
.200	.255	.700	1.046	1.200	2.706	1.700	6.070
.210	.268	.710	1.070	1.210	2.753		
.220	.280	.720	1.093	1.220	2.800		
.230	.293	.730	1.116	1.230	2.850		
.240	.305	.740	1.140	1.240	2.900		
.250	.319	.750	1.160	1.250	2.950		
.260	.330	.760	1.180	1.260	3.002		
.270	.342	.770	1.200	1.270	3.055		
.280	.356	.780	1.220	1.280	3.107		
.290	.370	.790	1.240	1.290	3.160		
.300	.385	.800	1.260	1.300	3.220		
.310	.399	.810	1.283	1.310	3.280		
.320	.412	.820	1.306	1.320	3.340		
.330	.426	.830	1.330	1.330	3.400		
.340	.440	.840	1.353	1.340	3.460		
.350	.455	.850	1.376	1.350	3.520		
.360	.470	.860	1.400	1.360	3.580		
.370	.484	.870	1.430	1.370	3.640		
.380	.499	.880	1.460	1.380	3.700		
.390	.514	.890	1.490	1.390	3.760		
.400	.530	.900	1.520	1.400	3.820		
.410	.547	.910	1.550	1.410	3.880		
.420	.564	.920	1.580	1.420	3.940		
.430	.580	.930	1.610	1.430	4.000		
.440	.600	.940	1.640	1.440	4.065		
.450	.617	.950	1.670	1.450	4.130		
.460	.633	.960	1.703	1.460	4.200		
.470	.650	.970	1.736	1.470	4.270		
.480	.666	.980	1.770	1.480	4.340		
.490	.683	.990	1.810	1.490	4.410		

Cell Synchrony

It is sometimes desirable to have an entire population of yeast cells that are growing synchronously or are arrested at a unique point in the cell cycle. One way to obtain synchrony is to use temperature-sensitive mutants that arrest at specific point in the cell division cycle (*cdc* mutants). However, this requires constructing special strains for each experiment. An alternative approach is to use drugs or inhibitors to arrest the cells in the cell cycle.

PROCDEURE

α-factor

MATα cells will arrest at START in the cell cycle in response to the dodecapeptide mating pheromone α-factor that is commercially available (SIGMA T6901). Store the peptide at 1 mg/ml in PBS at –20°C. *MATα* cells produce a protease (Bar1p) that destroys α-factor; therefore, successful use of α-factor in cell synchrony experiments requires that you accommodate for Bar1p activity. The degree of synchrony can be determined by the unique pear-shaped (schmoo) morphology that α-factor arrested cells adopt. There are three approaches.

Dilute Concentration of Cells

This is useful when a small number of cells are needed (e.g., for microscopy).

1. Grow cells to 10^7 cells per ml in YPD.
2. Concentrate cells by centrifugation and wash twice with sterile H_2O to remove Bar1p.
3. Resuspend cells at 10^4 cells per ml in YPD and add α-factor at 2 µg/ml.
4. Grow the cells for two generation times.
5. Concentrate cells by centrifugation and determine the percentage of cells that have arrested.

Low pH

The activity of Bar1p is pH dependent and can be partially suppressed at low pH.

1. Grow cells to 10^7 cells per ml in YPD.

2. Concentrate cells by centrifugation and wash twice with sterile H_2O to remove Bar1p.

3. Resuspend cells at 10^6 cells per ml in YPD (pH 3.5) and add α-factor at 2 μg/ml.

4. Grow the cells for two generation times.

5. Determine the percentage of cells that have arrested.

bar1 Mutants

bar1 mutants lack the Bar1p protease and are very sensitive to pheromone, even at high cell densities.

1. Grow cells to 10^7 cells per ml in YPD.

2. Add α-factor to 50 ng/ml.

3. Grow the cells for two generation times.

4. Determine the percentage of cells that have arrested.

Reentering the Cell Cycle

To release cells from the α-factor-induced arrest, centrifuge the cells and wash twice with H_2O. Resuspend the cells in medium containing 50 μg/ml of pronase.

Hydroxyurea

Hydoxyurea is an inhibitor of the enzyme ribonucleotide reductase, and cells in the presence of the drug are unable to synthesize deoxyribonucleotides and therefore cannot complete DNA synthesis. Treating a culture with hydroxyurea results in a population of cells that arrest in the early stages of S phase. Cells arrested with hydroxyurea are large-budded and have a single undivided nucleus and a short mitotic spindle. The effect of the drug is readily reversible.

1. Grow cells to 10^7 cells per ml in YPD.

2. Add hydroxyurea, as a powder, to a final concentration of 0.2 M.

3. Grow the cells for two generation times.

4. Determine the percentage of cells that have arrested.

Reentering the Cell Cycle

To release cells from the hydroxyurea-induced arrest, centrifuge the cells and wash twice with H_2O. Resuspend the cells in the appropriate medium.

Nocodazole

Nocodazole is an inhibitor of microtubule assembly, and cells in the presence of the drug cannot complete mitosis. Treating a culture with nocodazole results in a population of cells that arrest in M phase. Cells arrested with nocodazole are large-budded and have a single undivided nucleus and no mitotic spindle. The nucleus is not located at the neck but is distributed randomly in the cell. The effect of the drug is readily reversible.

1. Grow cells to 10^7 cells per ml in YPD.
2. Add 10 μl of nocodazole per ml of cells from a stock solution of nocodazole that is 1.5 mg/ml in DMSO (final concentration 15 μg/ml nocodazole, 1% DMSO).
3. Grow the cells for two generation times.
4. Determine the percentage of cells that have arrested.

Reentering the Cell Cycle

To release cells from the nocodazole-induced arrest, centrifuge the cells and wash twice with H_2O. Resuspend the cells in the appropriate medium.

Stationary Phase

Cells exit the cell cycle when they pass into stationary phase (G0 arrest). When cells in stationary phase are diluted into fresh medium, they will re-enter the cell cycle synchronously; however, the degree of synchrony is dependent on prior growth conditions and the strains. The utility of this method must be tested carefully before use.

1. Inoculate cultures at approximately 10^4 cells per ml in YP medium containing 2% raffinose and no glucose. Grow the cells for 48 hours at 30°C with constant aeration by shaking at 300 rpm in a rotary shaker. Make sure the culture volume is 10% of the volume of the flask to assure optimal aeration.
2. Concentrate the cells by centrifugation and resuspend them at a concentration of 5×10^6 cells per ml in YPD medium. Make sure the culture volume is 10% of the volume of the flask to assure optimal aeration. Assay cell synchrony by sampling every 15 minutes and determining the percentage of cells with small buds. Approximately 80% of the cells should initiate budding within a 15-minute interval after a 60- to 90-minute lag.

Chromatin Immunoprecipitation

PROCEDURE

1. Grow the yeast cells to be assayed. For each sample, grow 50–100 ml of cells to an OD_{600} of approximately 1.0 in a small culture flask.

2. Treat the cells with formaldehyde to crosslink proteins and DNA. Add formaldehyde to the cells to a final concentration of 1% (37% formaldehyde at 1:36) and maintain the cells at room temperature for 10–120 minutes, inverting occasionally.

   Note: The optimal fixation time varies depending on the protein of interest. This variable may need to be optimized for each specific assay.

3. Prepare lysis/IP buffer with protease inhibitors for the cell lysis step. I use 5–8 ml (more or less if desired; see steps 6 and 7) of lysis/IP buffer to collect each CHIP sample. The protease inhibitor stock solutions are 50x, so add 20 µl of each protease inhibitor per ml of lysis/IP buffer. Store the buffer in the cold room.

   Note: It is best to add PMSF last, *just before* using the buffer, since it is unstable in aqueous solutions, with a half-life of ~35 minutes at pH 8. The stocks are 50x, so add 20 µl of each protease inhibitor for each ml of buffer.

4. Quench the crosslinking reaction. Add 2.5 M glycine to a final concentration of 125 mM (a 20x dilution), and incubate the samples for 5 minutes at room temperature to quench the crosslinking reaction.

5. Wash the cells three times with TBS. Transfer the samples to GSA bottles or 50-ml polypropylene tubes to collect and wash the cells. Wash three times with ice-cold TBS, using ~20 ml of TBS twice, and less in the final wash. The final wash should be performed in 50-ml tubes, removing the TBS as thoroughly as possible in preparation for the lysis step.

6. Lyse the cells with glass beads.

   a. Resuspend the cells in lysis/IP buffer. Resuspend the cells in 250 µl of ice-cold lysis/IP buffer in the 50-ml tube.

   b. Add glass beads (0.5 mm) and vortex the cells. Add 3–6 ml of glass beads (for 1×10^9–2×10^9 cells) to the cells in lysis/IP buffer and use the S/P vortex mixer (it is the strongest machine) on the highest setting. After initially mixing the lysate and beads together, add enough beads so that there is ~1 ml of dry beads above the bead-lysate mixture in the tube. This may take an additional 1–2 ml of beads. The grinding action of the beads works best if there is relatively little liquid in relation to beads. Vortex each sample 6–8 times for 30 seconds each time or 4–5 times for 1 minute each time. It is best to do this step entirely in the cold room. If doing it at room temperature, keep the cells cold by placing them on ice periodically.

Note: Examine the cells under the microscope to check for effective cell lysis. If the lysis is incomplete, repeat the vortexing step as needed.

7. Collect the crude cell lysate. Add 2–3 ml of fresh lysis/IP buffer to each sample, and pipette off the crude cell lysate from the beads. Collect the crude lysate into a 15-ml snap-cap tube. To collect the lysate, use a 1-ml automatic pipettor, and put the pipette tip to the bottom of the tube before releasing. Perform this step three times (using the desired total volume of lysis buffer) to collect as much lysate as possible. The volume collected will be slightly less than the total added, since some stays with the beads. Do not worry about transferring a few glass beads, as these will be removed in subsequent steps.

Note: Take care not to cross-contaminate the lysate samples while pipetting or sonicating, since the final PCR analysis is very sensitive.

8. Sonicate the crude lysate to shear chromatin. Shear the sample chromatin by sonicating the suspension three times for 12–15 seconds each time. We use a Branson 250 sonifier with a microtip at power setting 3, 100% duty cycle. Between pulses, incubate the suspension on ice for at least 2 minutes. The average length of DNA postsonication should be 500 bp, with a range of 100–1000 bp.

Note: Check the DNA fragment sizes after sonication with your sonifier to be certain that you have small DNA fragments. To lower the chance of cross-contamination of samples you must always carefully clean the tip after each use of the sonifier.

9. Remove cell debris from the lysate by centrifugation. Spin the 15-ml tube at ~5K rpm for a few minutes in the cold room tabletop centrifuge to remove the bulk of the cell debris. Transfer the supernatant into a 15-ml Corex tube for further centrifugation steps. Pellet the cell debris at 10.75K in the Sorvall SS-34 rotor for 5 minutes at 4°C and decant the supernatant to a fresh tube. Repeat the centrifugation for 15 minutes at 4°C and again decant the supernatant to a fresh tube. The lysates should have a slightly milky appearance.

Note: One can stop at this point by freezing the lysate samples at –80°C.

10. Normalize the amounts of protein in the lysate samples. Perform a total protein quantitation assay for each lysate sample. Since the lysates are very concentrated, measure total protein in a 10x diluted aliquot. Adjust the concentration of the lysates by adding lysis/IP buffer as needed, and use equivalent amounts of protein in each IP.

11. Remove total chromatin samples. Remove a 50 μl sample of each cell lysate and add 200 μl of TE/1% SDS to it. These are the total chromatin samples. The samples should contain sheared genomic DNA, and are used to determine the equal presence of all DNA fragments prior to the immunoprecipitation. Do not IP these samples, but take them through the post-IP reversal of crosslinking and all subsequent steps to provide input DNA for the control PCR.

12. Immunoprecipitate the protein of choice (one- or two-step IP). Perform three-step IPs, using cell lysates with 30 mg of total protein in each IP.

    Preclearing. Preincubate the lysates with serum to remove nonspecific antigens. Use preimmune serum if available and "normal" serum if it is unavailable for this step.

    Primary antibody. Add an appropriate amount of antibody to the lysate, and incubate it for one to several hours at 4°C.

    No-antibody IP controls. Perform a no-antibody IP corresponding to each primary antibody IP, and follow it with an agarose bead IP identical to that for the other IP samples. This control will indicate the level of nonspecific immunoprecipitation due to the beads or other factors.

    Secondary antibody. Add 40 µl of a 50% suspension of protein A (for rabbit or rat antibodies) or protein G-agarose beads (for mouse antibodies). Incubate the samples at 4°C for 1 hour. This is a high-affinity binding step, and extended time is not likely to improve the IP.

13. Prepare lysis buffers with protease inhibitors for the post-IP washes. The post-IP wash step will require 2 ml of lysis buffer and 1 ml of lysis buffer/500 mM NaCl per sample. The protease inhibitor stock solutions are 50x, so add 20 µl of each protease inhibitor for each ml of buffer.

14. Wash the IP beads.

    a. Remove the IP supernatant. Pellet the beads for 1 minute at 3000 rpm in an microcentrifuge (~1000g), and remove the supernatant. It may be useful to save this fraction for later analysis.

    b. Wash the IP beads. Wash the beads according to the following schedule. Between washes, spin the beads down for 1 minute at 1000g (as in step 13), and remove the wash liquid with a small pipette tip attached to an aspirator, taking care to avoid the pellet. The washes should be performed at room temperature with 1 ml of the indicated solution, for 3–5 minutes each time.

       1. 2x with lysis buffer

       2. 1x with lysis buffer/500mM NaCl

       3. 2x with IP wash solution

       4. 1x with TE

15. Elute the immunoprecipitate with TES (TE/1% SDS).

    a. Elute the immunoprecipitate from the antibody beads with 110 µl of TES (pH 8.0), incubating at 65°C for 15 minutes.

    b. Pellet the beads for a few seconds at full speed (13,000 rpm) and transfer the eluate to a fresh tube. This is the first eluate fraction.

    c. Wash the beads with 150 µl of TE/0.67% SDS, mix them, and pellet again. Remove the supernatant, which is a second eluate fraction, and add it to the first fraction.

d. Spin the pooled eluate once more to eliminate any residual beads, and transfer it to a fresh tube. After centrifuging the combined eluate, remove 250 μl and avoid the ~10 μl left with the beads at the bottom of the tube.

16. Incubate all samples to reverse crosslinking. Incubate the eluates and the total chromatin samples at 68°C for at least 6 hours to reverse the crosslinking.

17. Treat the samples with proteinase K.

a. Add TE to the samples. After the crosslinking has been reversed, add 250 μl of TE to get a total sample volume of 500 μl.

b. Add glycogen and proteinase K. Add glycogen, a carrier for the DNA (add 5–10 μl of 10 mg/ml glycogen), and 100 μg of proteinase K (5 μl of 20 mg/ml stock) to each sample.

c. Incubate the samples. Incubate the samples for at least 2 hours at 37°C.

18. Add LiCl to the samples. Add 55 μl of 4 M LiCl to the solution. This salt will cause the DNA to precipitate in step 22.

19. Phenol/chloroform-extract the sample. Extract the samples with PCI (phenol:chloroform:isoamyl alcohol at 25:24:1). Extract the IP samples once with PCI followed by a chloroform:isoamyl alcohol extraction.

20. Precipitate the DNA.

a. Add ethanol to the samples. Add 1 ml of ethanol to each sample and mix. Store the samples in a cold place (I put them at –20°C or –70°C) to promote precipitation. Allow at least 15 minutes.

b. Centrifuge the samples. Centrifuge the DNA at 0°C. Centrifugation at 12,000g for 10 minutes is usually sufficient to pellet DNA, but DNA of low concentration may require longer. Rinse the pellet with 75% ethanol, spin down and remove the rinse liquid, and allow the pellet to air-dry.

21. RNase treatment. Resuspend the pellet in 25–50 μl of TE containing 10 μg of RNaseA. Incubate for 1 hour at 37°C.

22. Perform DNA amplification analysis of the sample DNA. Use specific primers to amplify DNA target. Use primers from a non-target locus as a negative control.

MATERIALS AND REAGENTS

Yeast cells to be assayed (1 x 10^9 to 2 x 10^9 cells per sample)
Formaldehyde (37%)
2.5 M glycine in H$_2$O
TBS (Tris-buffered saline)   1 liter 10x stock:
    20 mM Tris/HCl   200 ml of 1 M Tris/HCl, pH 7.6
    150 mM NaCl   300 ml of 5 M NaCl
                                         H$_2$O to reach 1 liter

NOTE:

Dilute to working concentration and store in the cold room, as it is to be used cold.

Lysis/IP buffer	500 ml:
50 mM HEPES/KOH	25 ml of 1 M HEPES/KOH, pH 7.5
140 mM NaCl	14 ml of 5 M NaCl
1 mM EDTA	1 ml of 500 mM EDTA
1% Triton X-100	50 ml of 10% Triton X-100
0.1% Na-deoxycholate	0.50 g Na-deoxycholate
	H$_2$O to reach 500 ml

Protease inhibitors	Stock solutions:
1 mM PMSF	50 mM PMSF in isopropanol
1 mM benzamidine	50 mM benzamidine in H$_2$O
1 mg/ml bacitracin	50 mg/ml bacitracin in H$_2$O

Store these stock solutions in small aliquots (~0.5–1 ml) at –20°C. Each ml of lysis/IP buffer or lysis buffer/500 mM NaCl to be used will require 20 μl of each of these protease inhibitor stock solutions.

Lysis buffer/500 mM NaCl	250 ml:
50 mM Hepes/KOH	12.5 ml of 1 M HEPES/KOH, pH 7.5
500 mM NaCl	25 ml of 5M NaCl
1 mM EDTA	0.5 ml 500 mM EDTA, pH 7.5
1% Triton X-100	25 ml of 10% Triton X-100
0.1% Na-deoxycholate	0.25 g Na-deoxycholate

NOTE:

Immediately before use, add protease inhibitors as for the lysis/IP buffer.

IP wash solution	250 ml:
10 mM Tris-HCl	2.5 ml of 1 M Tris-HCl, pH 8.0
0.25 M LiCl	12.5 ml of 5 M LiCl
0.5% NP-40	6.25 ml of 20% NP-40
0.5% Na-deoxycholate	1.25 g of Na-deoxycholate
1 mM EDTA	0.5 ml of 500 mM EDTA

TES (TE /1% SDS)	100 ml:
50 mM Tris/HCl	5 ml 1 M Tris/HCl, pH 8.0
10 mM EDTA	2 ml of 500 mM EDTA
1% SDS	5 ml 20% SDS
	H$_2$O to reach 100 ml

TE/0.67% SDS

    50 mM Tris-HCl

    10 mM EDTA

    0.67% SDS

100 ml:

    5 ml of 1 M Tris-HCl, pH 8.0

    2 ml of 500 mM EDTA

    3.35 ml of 20% SDS

    H_2O to reach 100 ml

TE (pH 8.0)

Branson Sonifier 250

primary antibody

protein A or G-agarose beads

Protein A-sepharose bead buffer

50 ml:

    TE, pH 7.5

    0.1% BSA

    0.1% sodium azide

47 ml of TE, pH 7.5

2.5 ml of 2% BSA

0.5 ml of 10% sodium azide

glycogen, 10 mg/ml in H_2O

proteinase K, 20 mg/ml in H_2O

Store the stock solution in 110-µl aliquots at –20°C. 100 µl of proteinase K will treat 20 samples (see step 16).

4 M LiCl in H_2O

PCI (phenol:chloroform:isoamyl alcohol, 25:24:1)

chloroform:isoamyl alcohol (24:1)

100% ethanol

75% ethanol/25% H_2O

RNase A

PCR reagents

Flow Cytometry of Yeast DNA

PROCEDURE

1. Grow and harvest the cells. Grow cells to 5×10^6 cells per ml. To collect the sample, spin down 10 ml of cells and wash them once in 5 ml of Tris buffer. Spin the cells down again and resuspend them in 1.5 ml of H_2O.

2. Fix the cells with ethanol. Add 3.5 ml of 100% ethanol to each 1.5-ml sample of cells to reach a final ethanol concentration of 70%. Leave the sample at room temperature for 1 hour. If necessary, store the sample at 4°C.

3. Wash out the ethanol and sonicate the cells. Spin the cells down and wash them in 5 ml of Tris buffer. Spin the cells down again, resuspend them in 5 ml of Tris buffer. Sonicate the sample to separate the cells (3x for 5–10 seconds each time).

4. RNase-treat the cells. Dilute the RNase stock soution to 1 mg/ml (1x) in Tris buffer. Spin the cells down, remove the buffer, and resuspend them in 2 ml of 1x RNase solution. Incubate the sample at 37°C for one hour with shaking, then overnight at 4°C if necessary. Complete RNase digestion is critical. This should be checked by looking at the cells with a fluorescence microscope (use rhodamine filters). The nuclear staining should be very bright and the cytoplasm should be devoid of stain except for mitochondrial DNA.

5. Treat the sample with pepsin. Spin down the cells and resuspend them in 1–2 ml of freshly prepared pepsin solution. Incubate the sample at room temperature for 5 minutes.

6. Stain the cells with propidium iodide. Spin down the cells and resuspend them in 2 ml of PI solution. Incubate for 1 hour at room temperature. Check the cells for proper staining by microscopy. Store the cells at 4°C in the dark and proceed with the analysis as soon as possible.

MATERIALS AND REAGENTS

Tris buffer
50 mM Tris/HCl (pH 7.5)

10x RNase solution
   10 mg/ml RNase
   100 mM NaOAc

10 ml of 10x stock:
   100 mg of RNase
   333 μl of 3 M of NaOAc (pH 5.0)
   9.7 ml of H_2O

Note: Boil the RNase stock solution for 30–60 minutes, and store it at –20°C. To use the RNase, dilute it 10x in Tris buffer. Use RNase type 1-A, 5x crystallized.

Pepsin solution

10 ml:
   50 mg of pepsin
   9.45 ml of H_2O
   550 μl of 1N HCl

Dissolve the pepsin in H_2O before adding the HCl. Use only freshly prepared pepsin solution.

100x propidium iodide stock solution
5 mg/ml propidium iodide in staining buffer

Store the propidium iodide stock solution in the dark at 4°C. For use, dilute the stock solution 100x in staining buffer (final concentration 50 μg/ml).

Staining buffer
   Tris 180 mM
   NaCl 180 mM
   $MgCl_2 \cdot 6H_2O$ 70 mM

1 liter:
   Tris 21.8 g
   NaCl 10.52 g
   $MgCl_2 \cdot 6H_2O$ 14.26 g
   adjust to pH 7.5

Logarithmic Growth

It is often desirable to perform experiments on yeast cells that are in all possible stages of the cell cycle. It is necessary to have cells in the mid-log stage of growth and not a culture with cells at stationary phase. Yeast cells grow with doubling times in the range of hours, and cells at stationary phase do not enter the cell cycle synchronously, except under carefully controlled growth conditions. Therefore, it is sometimes difficult and time-consuming to inoculate cells in the morning and be assured that they are normally distributed in the cell cycle. It is often necessary to inoculate cultures the day before to assure that they are at the right stage of growth for your use.

Logarithmic growth of a population of cells can be described mathematically as:

$$N = N_0 e^{\ln 2 \, (t/t_2)}$$

Where:

N = number of cells per ml at time = t

N_0 = the number of cells per ml at $t = 0$

t = time in hours

t_2 = doubling time in hours

The cell density of wild-type yeast cells grown to stationary phase in YPD is ~2×10^8 cells/ml and the cell density of wild-type yeast cells grown to stationary phase in SC or HC medium is ~2×10^7 cells/ml.

Doubling times are ~2 hours at 23°C, ~1.5 hours at 30°C, and ~1 hour at 36°C.

Example:

It is 5:00 PM. You want to start an experiment at 9:00 AM tomorrow using cells at a density of 1×10^7 per ml at 30°C. You have a 5-ml culture of cells at saturation in YPD. How much do you have to dilute the culture to get the cells at the right density in the morning?

$N = 1 \times 10^7$

$t = 16$ hours

$t_2 = 1.5$ hours at 30°C

$1 \times 10^7 = No \, e^{\ln 2 \, (16/1.5)}$

$1 \times 10^7 = No\ e^{(.69)(10.7)}$

$No = 1 \times 10^7/1608$

$No = 6218 = 6.2 \times 10^3$ cells per ml is the starting cell density that you want.

Dilute the cells: $2 \times 10^8/6218 = 3.2 \times 10^4$

Therefore, 3.2×10^4 is the fold dilution of the saturated culture that is necessary to get 6.2×10^3 cells per ml.

For a 50-ml culture (50,000 μl), you need to inoculate with 1.6 xml of the overnight culture; i.e., (50,000/32,000=1.6).

EMS Mutagenesis

PROCEDURE

SAFETY NOTES

Ethyl methanesulfonate (EMS) is a potent mutagen. Wear gloves and work in a hood when tubes are open. Use disposable pipettes for all manipulations and dispose of them in an appropriate waste container.

1. Grow an overnight culture to about 2×10^8 cells/ml.

2. Transfer two separate 1-ml samples of the same overnight culture to sterile microfuge tubes. Pellet the cells in a microfuge (10 seconds at $5000g$).

3. Discard the supernatant, resuspend the pellet in sterile distilled H_2O and re-pellet. Repeat.

4. Resuspend the cells in 1 ml of sterile 0.1 M sodium phosphate buffer (pH 7).

5. Determine the exact cell density with a counting chamber.

6. Add 30 µl of ethyl methanesulfonate (EMS) to one of the two tubes and disperse by vortexing vigorously. (The other tube will be the unmutagenized control.) Incubate both tubes for 1 hour at 30°C, with agitation.

7. Pellet the cells and remove the supernatant to a designated EMS waste container, then resuspend the cells in 200 µl of 5% sodium thiosulfate. Transfer to fresh tubes (discard the used tube in the EMS waste container).

8. Wash the cells twice with 200 µl of 5% sodium thiosulfate (each time discarding the supernatant in the EMS waste container). Resuspend the cells in 1 ml of sterile H_2O.

9. The cells may be plated directly, but in some cases it is important to provide a period of growth so that wild-type proteins in the cells can be replaced by the mutated versions, prior to exposing the cells to selective conditions. If outgrowth is performed, it is important to remember that this treatment results in the production of identical siblings in the culture.

NOTES:

This protocol will cause about 40–70% cell death in most haploid laboratory strains, but there is strain-dependent variation in sensitivity to the mutagen. This level of cell killing is commonly used in mutant hunts with haploid strains, but is not appropriate for all applications. Cells that survive this level of mutagenesis typically experience a 10^2–10^3 increase in the incidence of mutation relative to unmutagenized control cells (Lindegren et al. 1965). It is advisable to perform a pilot experiment with your strain to calibrate EMS-induced killing and mutation frequency. There are a number of simple tests that can be performed to assay mutation frequency. Perhaps the most useful is an assay for loss-of-function of the *CAN1* gene, which renders cells resistant to canavinine (Whelan et al. 1979). The generation of Can^r cells as a function of exposure to EMS can be monitored by plating mutagenized cells (along with unmutagenized control cells) on the canavinine-containing medium described in Appendix A. Another assay that monitors inactivation of a specific gene is the generation of cycloheximide-resistant cells (Kaufer et al. 1983). Cycloheximide resistance is nearly always the consequence of a specific transversion mutation in the *CYH2* gene. The assay is of limited utility for mutagens (such as EMS) that do not predominantly cause this type of mutation. Other assays that are commonly used to monitor mutagenesis include those in which inactivation of any of a few genes can yield the mutant phenotype. These include the generation of 5-FOA-resistant cells (loss of function of *URA3*, *URA5*, or other genes that may be involved in permeability; Boeke et al. 1986), growth on medium in which α-amino adipate is the sole nitrogen source (loss of *LYS2* or *LYS5* function; Chatto et al. 1979), and the generation of red colonies (loss of *ADE1* or *ADE2* function; Jones and Fink 1982).

REFERENCES

Boeke J.D., LaCroute F., and Fink G.R. 1986. A positive selection for mutants lacking orotidine-5´-phosphate decarboxylase activity in yeast: 5'-fluoro-orotic acid resistance. *Mol. Gen. Genet.* **197:** 345–346.

Chattoo B.B., Sherman F., Azubalis D.A., Fjellstedt T.A., Mehnert D., and Ogur M. 1979. Selection of *lys2* mutants of the yeast *Saccharomyces cerevisiae* by the utilization of α-aminoadipate. *Genetics* **93:** 51–65.

Jones E.W. and Fink G.R. 1982. Regulation of amino acid and nucleotide biosynthesis in yeast. In *The molecular biology of the yeast* Saccharomyces: *Metabolism and gene expression* (ed. J.N. Strathern et al.). Cold Spring Harbor Laboratory, Cold Spring Harbor, New York.

Kaufer N.F., Fried H.M., Schwindinger W.F., Jasin M., and Warner J.R. 1983. Cycloheximide resistance in yeast: The gene and its protein. *Nucleic Acids Res.* **11:** 3123–3135.

Lindegren G., Hwang L.Y., Oshima Y., and Lindegren C. 1965. Genetical mutants induced by ethyl methanesulfonate in *Saccharomyces*. *Can. J. Genet. Cytol.* **7:** 491–499.

Whelan W.L., Gocke E., and Manney T.R. 1979. The *CAN1* locus of *Saccharomyces cerevisiae:* Fine-structure analysis and forward mutation rates. *Genetics* **91:** 35–51.

Tetrad Dissection

PROCEDURE

1. Sporulate cells on either plates or liquid medium. Examine the sporulated cultures to confirm that tetrads have been produced. Cultures showing less than 5% tetrads are difficult to dissect.

2. Prepare a fresh solution of Zymolyase T100 (ICN) (0.05 mg/ml in 1 M sorbitol) and place it at 4°C. Zymolyase contains β-glucuronidase, which cleaves bonds in the ascus coat, making it easier to break apart the ascospores. A less expensive alternative to Zymolyase is Gluculase (typically used as a 1:10 dilution of the stock), which generally yields less efficient digestion of the ascus coat, with greater spheroplasting of ascospores, than does Zymolyase.

3. If sporulation was performed in liquid, (at a density of approximately 5×10^7 cells per ml), add 500 μl of the culture to a microfuge tube, centrifuge at 5000g for 10 seconds, and remove the supernatant. Gently resuspend the cell pellet in 50 μl of the Zymolyase solution and incubate at 30°C. If the culture was sporulated on plates, use the flat end of a sterile toothpick to transfer a dab of cells from the plate to a microfuge tube containing 50 μl of the Zymolyase solution. Suspend the cells by rotating the toothpick and incubate at 30°C.

3. Stop the Zymolyase digestion by placing the microfuge tube on ice and gently adding 150 μl of sterile H_2O. For most strains, an incubation time of approximately 10 minutes is appropriate, but some strains require significantly longer or shorter incubations. The appropriate incubation time can also vary for a specific strain as a function of sporulation time and conditions (e.g., liquid versus solid sporulation medium). If you have not previously dissected your strain, a good strategy is to remove samples from the Zymolyase digestion at timed intervals (2 minutes through 20 minutes). These samples can be examined microscopically to determine the ideal incubation time.

4. Very gently apply 10 μl of Zymolyase-treated cells as a streak across a YPD plate using a sterile loop. Alternatively, use a pipettor to transfer the cells to a tilted plate, allowing the droplet of cell suspension to run down the sloped agar surface leaving a stripe of Zymolyase-treated cells. With either method, care must be taken not to disrupt the tetrads that are now only tenuously intact due to the Zymolyase treatment. Generally, the stripe of Zymolyase-treated cells is made across the top or center of the plate. The cells must be accessible to the micromanipulator, and there must be sufficient space either above or below the stripe for placement of isolated ascospores.

5. Place the YPD plate containing the digested asci on the microscope stage. Be extremely careful not to break the microneedle. Adjust the stage so that a region of the plate without cells is in line with the objective lens. Focus the microscope on the surface of the plate and adjust the micromanipulator so the needle appears in the center of the field. Clean the needle by dragging it through the agar.

6. Move the stage so that the cells are visible. Look for clusters of four ascospores that have a clear zone around them. Pick up the four spores with the microneedle and place them on the agar at least 5 mm from the stripe of Zymolyase-treated cells. Note the position on the mechanical stage. Pick up three spores and move the stage 5 mm away from the streak. Deposit the three spores and pick up two spores. Move the stage 5 mm away from the streak, deposit the two spores, and pick up one spore. Move the stage 5 mm and deposit the remaining spore. Move the microscope stage to the left or right 5 mm from the line of the four spores and select another four-spore cluster. Separate the spores as before by 5-mm intervals. By this method, ten tetrads can be dissected on each side of the YPD plate. Remove the YPD plate from the stage, taking care not to break the microneedle.

7. Incubate the plate for 2–3 days at 30°C.

DISSECTING TIPS

1. In most sporulated cultures there are many nonsporulated cells. Care must be taken to select only true tetrads and not groups of four cells that may resemble a tetrad. Spores are very spherical and slightly refractory. Nonsporulated cells may be less round and are often larger than spores. Select groups of four in which all four cells are in direct contact with each other. Avoid groups of four that fall apart extremely easily. The spores in true tetrads are usually mildly cohesive, even after Zymolyase treatment. Never attempt to guess which cells belong to the tetrad if you accidentally pick up more than four cells.

2. To separate spores that do not come apart easily in the dissection process, place the needle on the agar surface with its edge against the spores. Tap the side of the microscope gently with your finger. This should cause the needle to vibrate and often results in the separation of the spores.

3. If a spore is lost or becomes inextricably embedded in the agar, use a marking pen to indicate the position of that tetrad on the plate before moving on to the next tetrad. This can be done by rotating the objective lens to the side and making a mark on the plate corresponding to the position of the tetrad in question.

4. If the tetrad you would like to dissect is so crowded by other cells that you cannot pick it up without picking up other cells, create a clear zone around your tetrad by moving the surrounding cells with your needle. Once a clear zone has been created, remove all cells that might be attached to the needle by dragging it through the agar in a region of the plate that is free of cells. Then return to your tetrad and transfer it to a clear area for dissection.

5. Plates for dissection should be level, clear, thin, and dry. Pour the plates on a level surface so that if one side of the plate is in focus, the other will be as well. Certain brands of agar contain considerable debris, some of which resembles yeast spores. Bacto agar from Difco is one brand that is usually free of this type of debris. Since dissection microscopes usually focus on cells through the agar, thinner plates give clearer resolution of cells. Plates containing 25 ml of medium work well for most tetrad dissection experiments.

Cells are picked up on dissection needles in a small water droplet. For this reason, it is very difficult to pick up cells from a fresh, wet plate. Dry plates for 2 days at room temperature or overnight in a 37°C incubator prior to using them as dissection plates.

Making a Tetrad Dissection Needle

Microneedles attached to any of a variety of types of micromanipulators are useful in the dissection of tetrads, isolation of zygotes from populations of mating haploid cells, and manipulation of individual cells. Microneedles can be purchased from commercial sources (Cora Styles Needles 'N Blocks [www.tiac.net/users/cstyles/]). Alternatively, it is possible to make needles by gluing a thin glass filament with a flat end to a bent glass capillary pipet. Two methods for making your own needles are described below.

Method 1. Microneedles be made by drawing thin filaments from a 2-mm-diameter glass rod using a small gas flame (Scott and Snow 1978). The exact diameter is not critical, and various investigators have different preferences. Spores are more readily picked up and transferred with microneedles having tips of larger diameters (100 μm), whereas manipulations in crowded areas having high densities of cells are more manageable with microneedles having smaller diameters (25 μm). A needle with a diameter of about 50 μm is an acceptable compromise. To draw out the needle, hold the glass rod over a low Bunsen burner flame until it is glowing orange. Simultaneously remove the rod from the flame while pulling the ends apart. With practice, it is possible to produce a glass thread using this method. Segments that appear to be the correct diameter are chopped into lengths of about 2 cm, placed on a glass slide that has been pre-wetted with water or saliva, then cut with a razor blade or glass coverslip to create microneedles 1 cM in length. The goal is create microneedles with one perfectly smooth flat end as illustrated in Figure 1. The short segments are inspected with a microscope to determine whether they have the correct diameter and a flat end.

The microneedle mounting rod is made from a 2-mm-diameter glass. A 100 μl capillary pipette works well for this. Heat the glass rod over a low flame about 1 cM from its end, and when it becomes pliable, bend it to a right angle. A drop of Super Glue is applied to this end of the mounting rod, which is then touched to one of the segments of a glass filament as shown in Figure 1. The filament is carefully positioned so that it is at a right angle to the axis of the mounting rod and so that it has the correct length. The length of the perpendicular end should be compatible with the distance between the needle holder of the micromanipulator being used and the surface of the dissection plate. A too-short needle will not reach the surface of the medium, and one that is too long will dig into the surface of the medium. The microneedle holder is fitted into the micromanipulator after the glue has dried.

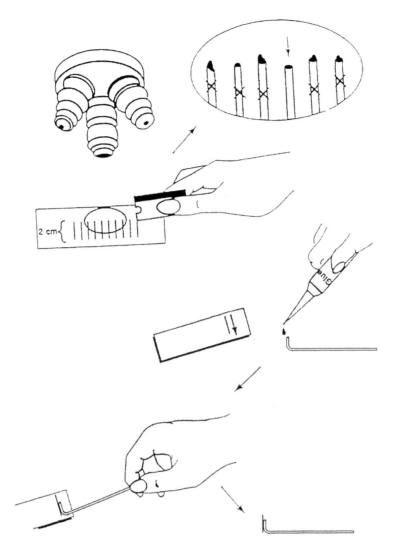

Figure 1. Making a dissection needle from drawn glass or optical fiber.

Method 2. This method is the same as Method 1 except that instead of pulling out a needle, commercially available fiber-optic glass (Edmond Scientific) of the correct diameter is used. The glass fiber is cut with a scissors to produce pieces that can be placed on a glass slide. These can then be cut with a razor blade or glass coverslip as described in METHOD 1 to create microneedles with flat ends. As described above, acceptable microneedles can then be mounted on a glass rod and attached to the micromanipulator.

REFERENCE

Scott K.E. and Snow R. 1978. A rapid method for making glass micromanipulator needles for use with microbial cells. *J. Gen. Appl. Microbiol.* **24:** 295–296.

Picking Zygotes

The unique morphology of zygotes makes it possible to identify them among populations of mating haploid cells. Zygotes can be easily separated away from nonmated cells using a micromanipulator. This method provides an alternative to the selection of diploid cells from a mating mixture by plating the mixture on a medium on which the diploid, but neither haploid parent, can grow.

Mating will be most efficient if the parent cells are from fresh cultures. Cells harvested from plates incubated 1–2 days at 30°C generally work well. Cells from older cultures including those stored for several days at 4°C will mate, but the formation of zygotes will take longer than it would for fresh cultures.

PROCEDURE

1. Use sterile toothpicks to place equal amounts of the *MAT*a and *MAT*α parent cells (about the size of a 1–2 mm sphere) adjacent to each other on the surface of a YPD plate, then mix them well, creating a circle of about 5 mm in diameter.

2. Incubate the plate at 30°C for 3 hours.

3. Use a sterile toothpick to streak a sample of the mating mixture across the top of a fresh plate, such that individual cells are distinguishable when the plate is observed using a tetrad-dissecting microscope.

4. Zygotes form through the fusion of shmoos, the elongated haploid cells that have arrested in response to mating pheromone. These zygotes have a dumbbell shape that can be confused with large budded haploid cells in the mating mixture. Zygotes that have formed a medial bud have a distinctive three lobed shape that allows them to be readily identified in the mating mixture.

5. Use methods described in Techniques & Protocols #21, Tetrad Dissection, to isolate these zygotes to a clean area of the plate using a microneedle designed for tetrad dissection.

6. Incubate the plate for 1–2 days at 30°C.

7. To confirm that the colonies arising from your isolated cells are indeed diploid, replica-plate them to media that will confirm the predicted phenotypes of your diploid strain. If the two haploid parents have no distinctive genetic markers, use mating test lawns (described in Experiment 2) to confirm that your strain is a non-mater, the predicted mating type of a *MAT*a/*MAT*α diploid.

Determining Plating Efficiency

The plating efficiency of a strain measures the percentage of viable cells in a culture that are capable of forming colonies (sometime called colony forming units or c.f.u.). There are two simple procedures that can be used to determine plating efficiency.

PROCEDURE

Indirect Method

1. Determine the cell density in your culture using a Coulter counter, by determining the optical density of the culture or by counting the cell number using a hemocytometer.

2. Determine the dilution factor necessary to dilute your cells to 10^3 cells per ml. Sonicate the cells to disperse clumps and dilute the cells in sterile H_2O.

3. Plate 0.2 ml of the diluted cells on a YPD plate, incubate for several days at the required temperature, and count the number of colonies. Determine the plating efficiency as the number of colonies divided by 200. Express the number as a decimal (100 colonies is a plating efficiency of 0.5).

Direct Method

1. Use a culture of cells at 10^6–10^7 cells per ml. Sonicate to disperse cells.

2. Spread 0.2 ml of the cells onto the surface of a YPD plate and allow the liquid to dry on the plate.

3. Incubate the plate for 16–24 hours at the desired temperature.

4. Observe the surface of the plate with a tetrad dissection microscope. Viable cells will produce a small rounded microcolony of 50–100 cells. Inviable cells will not form a colony but will be a disorganized collection of 1–10 cells. Count the number of viable and inviable cells directly and determine the plating efficiency.

Media

Media for petri plates are prepared in 2-liter flasks, with each flask containing 1 liter of medium, which is sufficient for 30–40 plates. Unless otherwise stated, all components are autoclaved together for 15 minutes at 250°F (121°C) and 15 lb/sq. in. of pressure. Longer autoclaving of minimal media leads to hydrolysis of the agar, caramelized glucose, and mushy plates. If larger volumes are to be prepared, autoclave the salts, glucose, and agar separately for longer periods of time. The plates should be allowed to dry at room temperature for 2–3 days after pouring. The plates can be stored in sealed plastic bags for over 3 months. The agar is omitted for liquid media. (For convenience, the final concentration of each component in the medium is listed in parentheses below.)

YPD (YEPD)

YPD is a complex medium for routine growth.

Bacto-yeast extract (1%)	10 g
Bacto-peptone (2%)	20 g
Glucose (2%)	20 g
Bacto-agar (2%)	20 g
Distilled H_2O	1000 ml

YPG (YEPG OR YEP-GLYCEROL)

YPG is a complex medium containing a nonfermentable carbon source (glycerol) that does not support the growth of ρ^- or *pet* mutants.

Bacto-yeast extract (1%)	10 g
Bacto-peptone (2%)	20 g
Glycerol (3% [v/v])	30 ml
Bacto-agar (2%)	20 g
Distilled H_2O	970 ml

YPDG

YPDG is a complex medium that can be used to differentiate between ρ^+ and ρ^- colonies. Combine:

Bacto-agar (2%)	20 g
Bacto-yeast extract (1%)	10 g
Bacto-peptone (2%)	20 g
Distilled H_2O	900 ml

After autoclaving the above ingredients together, add sterile:

30% Glycerol (3%)	100 ml
20% Glucose (0.1%)	5 ml

YPAD (SLANT MEDIUM)

YPAD is a complex medium used for the preparation of slants. The adenine is added to inhibit the reversion of *ade1* and *ade2* mutants.

Bacto-yeast extract (1%)	10 g
Bacto-peptone (2%)	20 g
Glucose (2%)	20 g
Adenine sulfate (0.004%)	40 mg
Bacto-agar (2%)	20 g
Distilled H_2O	1000 ml

Dissolve the medium in a boiling-water bath. Dispense 1.5-ml portions with an automatic pipettor into 1-dram vials. Screw on the caps loosely and autoclave the vials. After autoclaving, incline the rack so that the agar is just below the neck of the vial. Tighten the caps after 1–2 days.

SYNTHETIC DEXTROSE MINIMAL MEDIUM (SD)

SD is a synthetic minimal medium containing salts, trace elements, vitamins, a nitrogen source (Bacto-yeast nitrogen base without amino acids), and glucose.

Bacto-yeast nitrogen base without amino acids (0.67%)	6.7 g
Glucose (2%)	20 g
Bacto-agar (2%)	20 g
Distilled H_2O	1000 ml

SUPPLEMENTED MINIMAL MEDIUM (SMM)

SMM is SD to which various growth supplements have been added. The specific constituents in SMM are defined in the Materials section of each experiment, where applicable. It is convenient to prepare sterile stock solutions by autoclaving for 15 minutes at 250°F (121°C). These solutions can then be stored for extensive periods. Some should be stored at room temperature in order to prevent precipitation, whereas the other solutions may be refrigerated. Wherever applicable, HCl salts of amino acids are preferred.

The medium should be prepared by adding the appropriate volumes of the stock solutions to the ingredients of SD medium and then adjusting the total volume to 1 liter with distilled H_2O. Threonine and aspartic acid solutions should be added separately to the medium after it is autoclaved.

Alternatively, it is often more convenient to prepare the medium by spreading a small quantity of the supplement(s) on the surface of an SD plate. The solution(s) should then be allowed to dry thoroughly onto the plate before inoculating it with yeast strains.

Given below are the concentrations of the stock solutions, the volume of stock solution necessary for mixing a liter of medium, and the volume of stock solution to spread on SD plates. The final concentration of each constituent in SMM is also given.

Constituent	Stock concentration (g/100 ml)	Volume of stock for 1 liter of medium (ml)	Final concentration in medium (mg/liter)	Volume of stock to spread on plate (ml)
Adenine sulfate	0.2[a]	10	20	0.2
Uracil	0.2[a]	10	20	0.2
L-Tryptophan	1	2	20	0.1
L-Histidine HCl	1	2	20	0.1
L-Arginine HCl	1	2	20	0.1
L-Methionine	1	2	20	0.1
L-Tyrosine	0.2	15	30	0.2
L-Leucine	1	10	100	0.1
L-Isoleucine	1	3	30	0.1
L-Lysine HCl	1	3	30	0.1
L-Phenylalanine	1[a]	5	50	0.1
L-Glutamic acid	1[a]	10	100	0.2
L-Aspartic acid	1[a,b]	10	100	0.2
L-Valine	3	5	150	0.1
L-Threonine	4[a,b]	5	200	0.1
L-Serine	8	5	400	0.1

[a]Store at room temperature.
[b]Add after autoclaving the medium.

SYNTHETIC COMPLETE (SC) AND DROPOUT MEDIA

In order to test the growth requirements of strains, it is useful to have media in which each of the commonly encountered auxotrophies is supplemented except the one of interest (dropout media). Dry growth supplements are stored premixed.

SC is a medium in which the dropout mix contains all possible supplements (i.e., nothing is "dropped out").

Bacto-yeast nitrogen base without amino acids (0.67%)	6.7 g
Glucose (2%)	20 g
Bacto-agar (2%)	20 g
Dropout mix (0.2%)	2 g
Distilled H_2O	1000 ml

Dropout mix:

Dropout mix is a combination of the following ingredients minus the appropriate supplement. It should be mixed very thoroughly by turning end-over-end for at least 15 minutes; adding a couple of clean marbles helps.

Adenine	0.5 g	Leucine	10.0 g
Alanine	2.0 g	Lysine	2.0 g
Arginine	2.0 g	Methionine	2.0 g
Asparagine	2.0 g	*para*-Aminobenzoic acid	2.0 g
Aspartic acid	2.0 g	Phenylalanine	2.0 g
Cysteine	2.0 g	Proline	2.0 g
Glutamine	2.0 g	Serine	2.0 g
Glutamic acid	2.0 g	Threonine	2.0 g
Glycine	2.0 g	Tryptophan	2.0 g
Histidine	2.0 g	Tyrosine	2.0 g
Inositol	2.0 g	Uracil	2.0 g
Isoleucine	2.0 g	Valine	2.0 g

HARTWELL'S COMPLETE (HC) MEDIUM

HC medium, used in Lee Hartwell's lab, is used in the same way as SC medium (see above); however, it uses different combinations of supplements for growth. This difference supports better growth of some strains. In addition, it is presented as a series of stock solutions, which gives very reproducible results between batches of media, and with a preference for easily making the six most common dropout media.

The stock solutions are:

10x HC *(6 Dropout amino acid liquid):*

Methionine	0.8 g
Tyrosine	2.4 g
Isoleucine	3.2 g
Phenylalanine	2.0 g
Glutamic acid	4.0 g
Threonine	8.0 g
Aspartic acid	4.0 g
Valine	6.0 g
Serine	16.0 g
Arginine	0.8 g

in a final volume of 4 liters; autoclave

10x YNB:

Yeast nitrogen base	58 g
(w/o amino acids and w/o ammonium sulfate)	
Ammonium sulfate	200 g

in a final volume of 4 liters; autoclave

Amino acid solutions (note that these are NOT 10x solutions; see below)

Sterilize each solution by autoclaving. Keep the tryptophan in a light-sensitive bottle.

Uracil	1 g/liter
Adenine	1 g/liter
Lysine	10 g/liter
Tryptophan	10 g/liter
Leucine	20 g/liter
Histidine	10 g/liter

Recipe for Hartwell Complete Plates:

20 g of agar in 619 ml of distilled H_2O in a 2-liter flask. Autoclave 20 minutes.   Then add the following:

20% Glucose (sterile)	100 ml
10x YNB solution	100 ml
10x HC dropout 6 amino acid solution	100 ml
Uracil solution	35 ml
Adenine solution	20 ml
Lysine solution	12 ml
Tryptophan solution	8 ml
Leucine solution	4 ml
Histidine solution	2 ml

MAL INDICATOR MEDIUM

MAL indicator medium is a fermentation-indicator medium used to distinguish strains that ferment or do not ferment maltose. Due to the pH change, the maltose-fermenting strains will change the indicator yellow.

Bacto-yeast extract (1%)	10 g
Bacto-peptone (2%)	20 g
Maltose (2%)	20 g
Bromcresol purple solution (0.4% stock solution)	9 ml
Bacto-agar (2%)	20 g
Distilled H_2O	1000 ml

0.4% Bromcresol purple solution:

Bromcresol purple	200 mg
100% Ethanol	50 ml

GAL INDICATOR MEDIUM

GAL indicator medium is used for scoring the ability to ferment galactose.

Bacto-yeast extract (1%)	10 g
Peptone (2%)	20 g
Bacto-agar (2%)	20 g
Bromthymol blue solution (4 mg/ml stock solution)	20 ml
Distilled H_2O	880 ml

After autoclaving, add 100 ml of a filter-sterilized (0.2-μm filter) 20% galactose solution.

Bromthymol blue solution:

Bromthymol blue	400 mg
Distilled H_2O	100 ml

X-GAL INDICATOR PLATES FOR YEAST

5-Bromo-4-chloro-3-indolyl-D-galactoside (X-gal) does not work for yeast at the normal acidic pH of SD medium; therefore, a neutral pH medium is used. This is clearly a trade-off as many yeast strains will not grow well at this pH. For a first attempt at assessing β-galactosidase expression, this medium is worth a shot.

For 1 liter of X-gal indicator plates:

Solution I:

Mix:

10x Phosphate-buffer stock solution	100 ml
1000x Mineral stock solution	1 ml
Dropout mix	2 g

Adjust the volume to 450 ml with distilled H_2O if the medium is to contain glucose, or to 400 ml if it is to contain galactose.

Solution II:

Mix in a 2-liter flask:

Bacto-agar	20 g
Distilled H_2O	500 ml

Autoclave the solutions separately. After cooling to below 65°C, add the following to Solution I:

Glucose or other sugar to a final concentration of 2%	
X-gal (20 mg/ml dissolved in dimethylformamide)	2 ml
100x Vitamin stock solution	10 ml
Any other heat-sensitive supplements	

Mix the solutions together and pour ~30 ml/plate.

10x Phosphate-buffer stock solution:

KH_2PO_4 (1 M)	136.1 g
$(NH_4)_2SO_4$ (0.15 M)	19.8 g
KOH (0.75 N)	42.1 g
Distilled H_2O	1000 ml

Adjust the pH to 7 and autoclave.

SAFETY NOTE

Solid KOH is caustic and should be handled with great care. Gloves and a face protector should be worn.

1000x Mineral stock solution:

$FeCl_3$ (2 mM)	32 mg
$MgSO_4 \cdot 7H_2O$ (0.8 M)	19.72 g
Distilled H_2O	100 ml

Autoclave and store. This solution will form a fine yellow precipitate, which should be resuspended before use.

100x Vitamin stock solution:

Thiamine (0.04 mg/ml)	4 mg
Biotin (2 μmg/ml)	0.2 mg
Pyridoxine (0.04 mg/ml)	4 mg
Inositol (0.2 mg/ml)	20 mg
Pantothenic acid (0.04 mg/ml)	4 mg
Distilled H_2O	100 ml

Filter-sterilize using a 0.2-μm filter.

X-GAL PLATES FOR LYSED YEAST CELLS ON FILTERS

These plates are used for checking β-galactosidase activity in cells that have been lysed and are immobilized on 3MM filters.

Bacto-agar	20 g
1 M Na_2HPO_4	57.7 ml
1 M NaH_2PO_4	42.3 ml
$MgSO_4$	0.25 g
Distilled H_2O	900 ml

After autoclaving, add 6 ml of X-Gal solution (20 mg/ml in *N,N*-dimethylformamide).

SPORULATION MEDIUM

Strains will undergo several divisions on this medium and then sporulate after 3–5 days of incubation.

Potassium acetate (1%)	10 g
Bacto-yeast extract (0.1%)	1 g
Glucose (0.05%)	0.5 g
Bacto-agar (2%)	20 g
Distilled H_2O	1000 ml

Nutritional supplements are required for sporulation of auxotrophic diploids on sporulation medium. Supplements at the level of 25% of those used for SMM plates should be added when compounding the medium. Alternatively, supplements can be spread on the surface of the sporulation plates in the volumes listed for SMM. The liquid should be allowed to dry thoroughly onto the agar before inoculating it with yeast strains.

MINIMAL SPORULATION MEDIUM

*MAT*a/*MAT*α diploid cells will sporulate on this medium after 18–24 hours without vegetative growth.

Potassium acetate (1%)	10 g
Bacto-agar (2%)	20 g
Distilled H_2O	1000 ml

Nutritional supplements are required for sporulation of auxotrophic diploids on sporulation medium. Supplements at the level of 25% of those used for SMM plates should be added when compounding the medium. Alternatively, supplements can be spread on the surface of the sporulation plates in the volumes listed for SMM. The liquid should be allowed to dry thoroughly onto the agar before inoculating it with yeast strains.

LOW-pH BLUE PLATES

These plates are used for testing killer phenotype.

Bacto-yeast extract (1%)	6 g
Bacto-peptone (2%)	12 g
Glucose (2%)	12 g
Bacto-agar (2%)	12 g
Distilled H_2O	533 ml

Autoclave the above ingredients and add the following solutions:

Methylene blue in sterile H_2O	5 ml
Phosphate-citrate buffer (sterile)	67 ml

Methylene blue in sterile H_2O:

Methylene blue	20 mg
Sterile H_2O	5 ml

Phosphate-citrate buffer for low-pH medium:

Citric acid	14.07 g
K_2HPO_4	18.96 g
Distilled H_2O	67 ml

Adjust the pH to 4.5 using solid K_2HPO_4 or citric acid. Sterilize by autoclaving.

DRUG SELECTION MEDIA

5-Fluoro-orotic Acid Medium

5-Fluoro-orotic acid (5-FOA) can be used to select for mutant cells that fail to utilize orotic acid as the source of the pyrimidine ring. Wild-type cells convert 5-FOA to 5-fluoro-orotidine monophosphate by conjugation to phosphoribosyl pyrophosphate (PRPP), and subsequently decarboxylate it to form 5-fluoro-uridine monophosphate (5-FUMP). These two steps are catalyzed by the products of the yeast genes *URA5* and *URA3*, respectively. Inevitably, fluorodeoxyuridine formed later is a potent inhibitor of thymidylate synthetase and thereby quite toxic to the cell. The two steps of de novo synthesis of uridine that are required to convert 5-FOA to 5-FUMP can be mutated to block utilization, as long as uracil is provided to allow formation of UMP via the salvage pathway. Therefore, both *ura3*⁻ and *ura5*⁻ mutants can grow on 5-FOA-containing medium (Boeke et al. 1984). In practice, only *ura3*⁻ mutants appear to be uracil auxotrophs. The enzyme that catalyzes conjugation of uracil to PRPP can utilize orotic acid as a substrate at some level allowing *ura5*⁻ mutants to grow slowly in the absence of uracil.

Bacto-yeast nitrogen base (0.67%) 6.7 g

Drop-out mix ⁻ ura (0.2%)   2 g

Glucose (2%)   20 g

Uracil (50 μg/ml)   50 mg

5-FOA (0.1%)   1 g

Distilled H$_2$O   500 ml

Dissolve the above and filter-sterilize using a 0.2-μm filter.

Autoclave the agar separately:

Bacto-agar (2%)   20 g

Distilled H$_2$O   500 ml

Mix the two solutions after cooling the agar to ~80°C. Pour into petri dishes (25 ml/dish).

5-FOA Medium ala HC

Make one liter of HC plates as described above, let flask cool to 50°C, then add 1.0 g of 5-FOA.

α-Aminoadipate Plates

Wild-type strains are unable to utilize high levels of α-aminoadipate (αAA) as their sole nitrogen source because it is converted into a toxic intermediate by the normal lysine anabolic pathway (Chattoo et al. 1979; Zaret and Sherman 1985). This medium is frequently used in the selection of mutations in the *LYS2* and *LYS5* genes.

    Bacto-yeast nitrogen base without amino acids

      or ammonium sulfate (0.16%)   1.6 g

    Glucose (2%)   20 g

    Lysine (30 mg/liter)   30 mg

    Bacto-agar (2%)   20 g

    Distilled H$_2$O   960 ml

    Autoclave and add 40 ml of a 5% αAA solution.

5% αAA:

    α-Aminoadipic acid   2 g

    Distilled H$_2$O   40 ml

    Mix and adjust the pH to 6 with 10 N KOH to allow dissolution. Filter-sterilize using a 0.2-μm filter before adding to the autoclaved ingredients.

SAFETY NOTE

    Concentrated bases should be handled with great care; gloves and a face protector should be worn.

Cycloheximide

Cycloheximide resistance can arise in a number of different genes, but resistance to high levels ordinarily occurs due to rare mutations at the *cyh2* locus, which encodes the L29 ribosomal subunit. Resistance to cycloheximide is recessive, presumably because the sensitive ribosomes remain bound to the mRNA and block all further elongation.

Cycloheximide can be used in either YPD or synthetic media. A final concentration of 10 mg/liter should be used for YPD and 3 mg/liter for SD, SC, and YPG. A stock solution is prepared by dissolving 100 mg of cycloheximide in 10 ml of distilled H_2O and then filter-sterilizing (0.2-μm filter). The stock solution can be stored at 4°C. Appropriate volumes can be added to media after autoclaving.

Canavanine

Canavanine is an analog of arginine. Both are imported into the cell via the same high-affinity permease, which is encoded by the *CAN1* locus. High-level resistance to canavanine occurs exclusively because of mutation at this locus, but low-level resistance can arise at a number of other loci.

Because canavanine is a competitive inhibitor, arginine must be excluded from media used for testing sensitivity to the drug. Canavanine resistance must also be scored under high-nitrogen conditions, such as those provided by SD or SC medium, since the *CAN1* permease will then provide the only entry route to the cell for arginine and canavanine. In the presence of low-nitrogen conditions—effectively those provided by YPD medium—the general amino acid permease (*GAP*) system is induced and arginine and canavanine can also be taken up by this route. In addition, Can^R Arg^- auxotrophs are viable on YPD but are inviable on synthetic media because they are unable to take up arginine.

Canavanine sulfate is typically made up as a filter-sterilized (0.2-μm filter) 20 mg/ml stock solution in distilled H_2O. It is stored at 4°C and added to SD or SC – arg medium after autoclaving. A concentration of 60 mg/liter is typically used for scoring and selecting canavanine resistance.

REFERENCES

Boeke J.D., LaCroute F., and Fink G.R. 1984. A positive selection for mutants lacking orotidine-5´-phosphate decarboxylase activity in yeast: 5-Fluoro-orotic acid resistance. *Mol. Gen. Genet.* **197:** 345–346.

Chattoo B.B., Sherman F., Azubalis D.A., Fjellstedt T.A., Mehnert D., and Ogur M. 1979. Selection of *lys*2 mutants of the yeast *Saccharomyces cerevisiae* by the utilization of α-aminoadipate. *Genetics* **93:** 51–65.

Zaret K.S. and Sherman F. 1985. α-Aminoadipate as a primary nitrogen source for *Saccharomyces cerevisiae* mutants. *J. Bacteriol.* **162:** 579–583.

Stock Preservation

Yeast strains can be stored indefinitely in 15% (v/v) glycerol at a temperature of –60°C or less (Well and Stewart 1973). Yeast tends to die if stored at temperatures above –55°C. Many workers use 2-ml vials (35- x 12-mm) containing 1 ml of sterile 15% (v/v) glycerol. The strains are grown on the surfaces of YPD plates. The yeast is then scraped up with sterile applicator sticks or toothpicks and suspended in the glycerol solution. The caps are tightened and the vials shaken before freezing. The yeast can be revived by transferring a small portion of the frozen sample onto a YPD plate.

Yeast strains can be stored at 4°C for up to 6 months on slants prepared with YPAD medium. This method of storage is convenient since the slants take up little space, do not dry out, and contain excess adenine to prevent toxicity due to the red pigment produced by certain *ade⁻* mutants. Slants are also a useful means of sending strains to colleagues.

REFERENCE

Well A.M. and Stewart G.G. 1973. Storage of brewing yeasts by liquid nitrogen refrigeration. *Appl. Microbiol.* **26:** 577.

Yeast Genetic and Physical Maps

Figures I–XVI on the following pages (186–191) depict the genetic and physical maps, and their correlations, of the 16 *Saccharomyces cerevisiae* chromosomes. A parallel comparison of the physical map (left, in kilobase pairs) and the genetic map (right, in centimorgans) of each of the 16 chromosomes is illustrated. The information in this figure is available on the *Saccharomyces* Genome Database:

(http://genome-www.stanford.edu/Saccharomyces/).

The physical map consists of shaded boxes that indicate ORFs. ORFs on the Watson strand (left telomere is the 5′ end of this strand) are shown as light gray boxes; those on the Crick strand as dark gray boxes. Where it has been defined, the gene name of an ORF is indicated. The genetic map is based on data collected since 1991 by the SGD project as well as earlier data. Horizonal tick marks on the right of the genetic map line indicate positions of genes. Lines connect genetically mapped genes with their ORF on the physical map. A single name is listed for known synonyms. (Reprinted, with permission, from Cherry et al. 1997 [©Macmillan Magazines Ltd.]).

REFERENCE

Cherry J.M., Ball C., Weng S., Juvik G., Schmidt R., Adler C., Dunn B., Dwight S., Riles L., Mortimer R.K., and Botstein D. 1997. Genetic and physical maps of *Saccharomyces cerevisiae*. *Nature* (suppl.) **387:** 67–73.

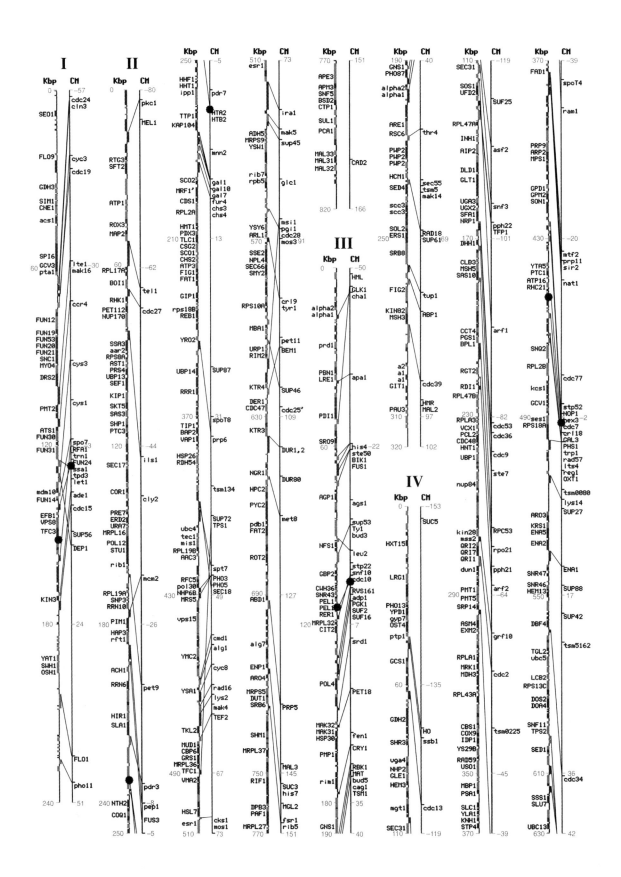

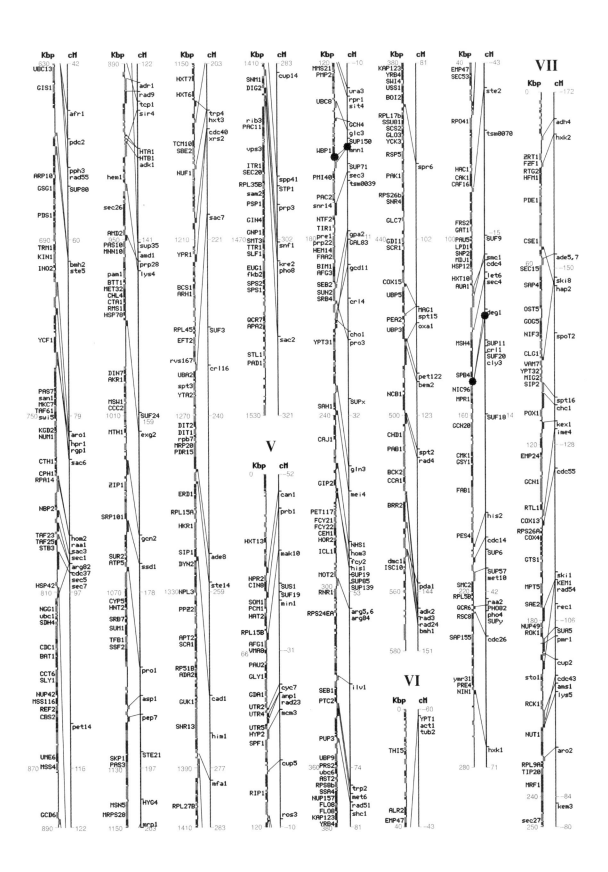

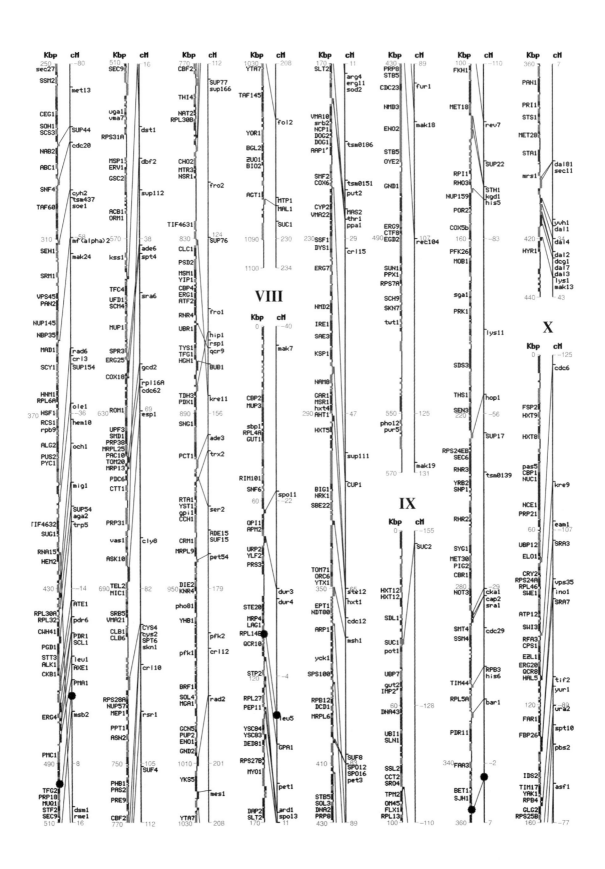

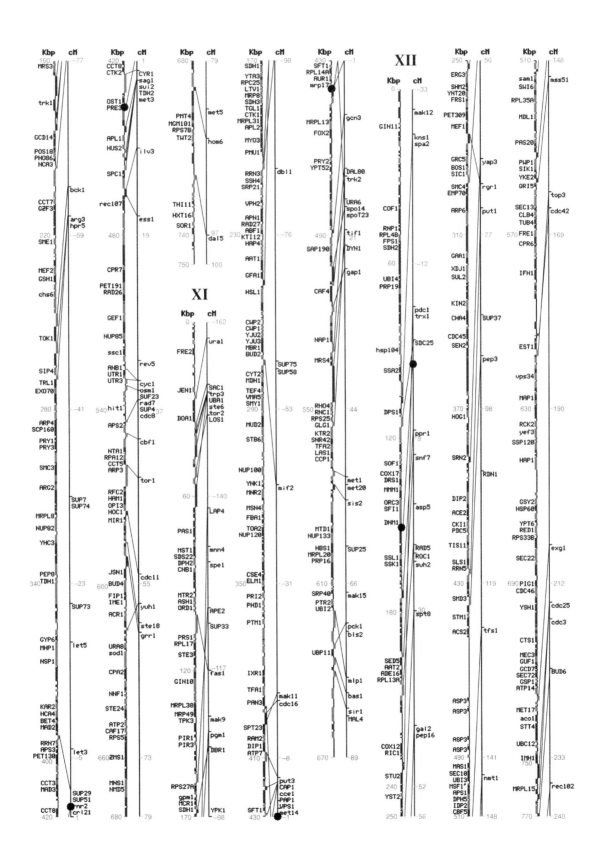

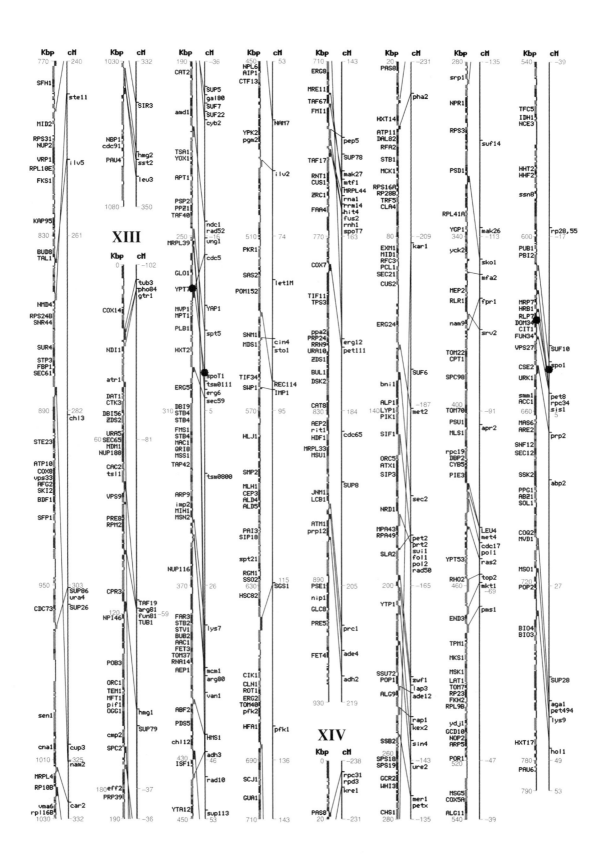

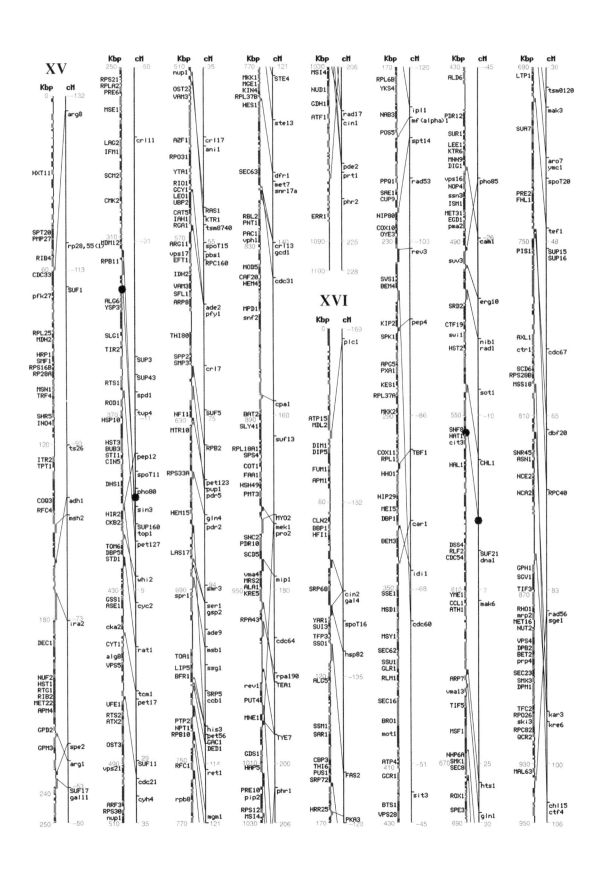

Templates for Making Streak Plates

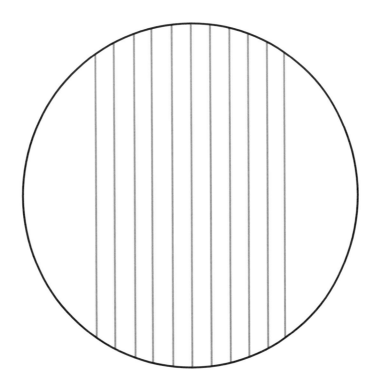

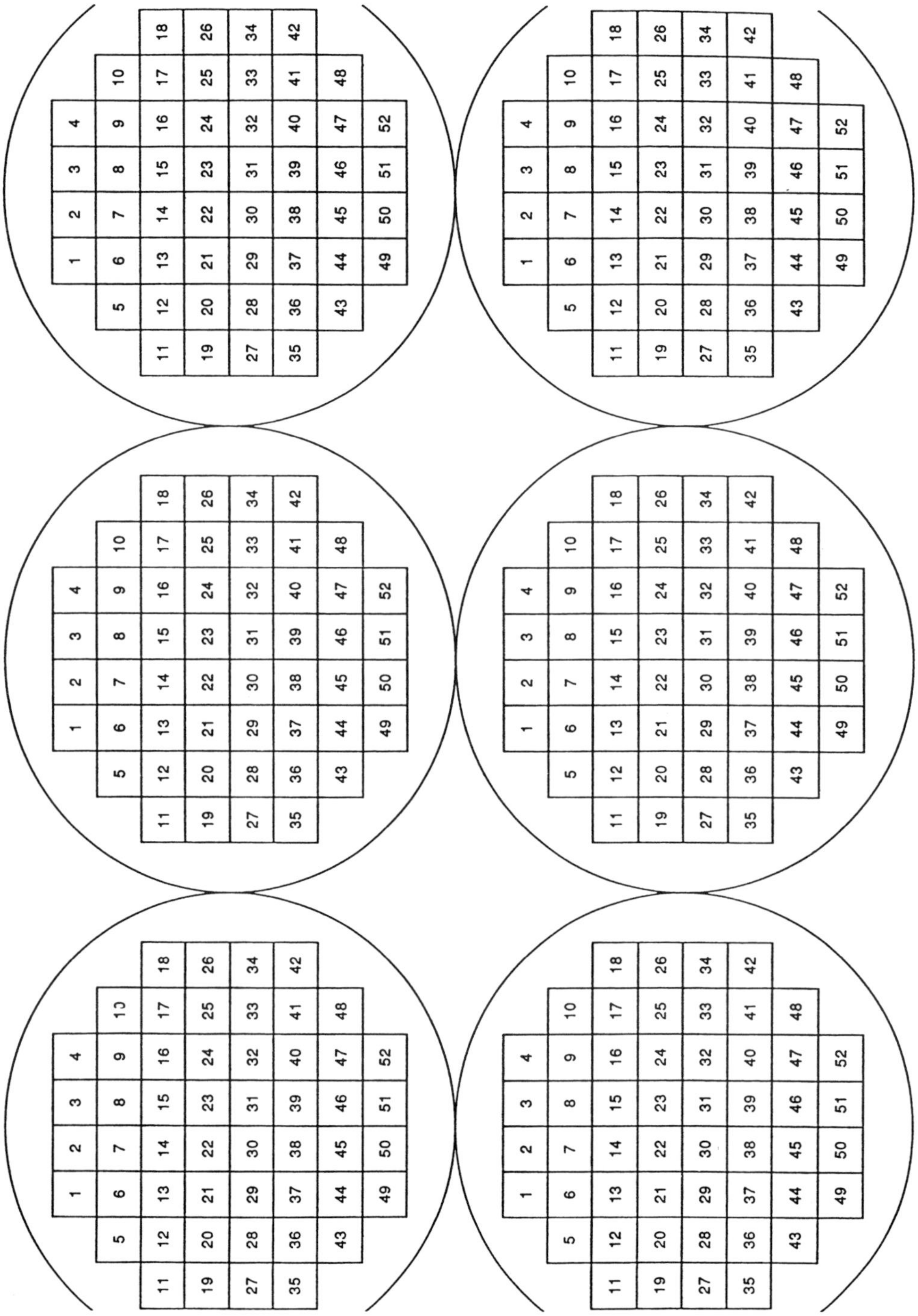

Electrophoretic Karyotypes of Strains for Southern Blot Mapping

NOTE:

This appendix has been largely superseded by the sequencing of the yeast genome (see Appendix C), but this information is still useful for evaluating the size of whole chromosomes by pulsed-field gel electrophoresis.

CHROMOSOME ASSIGNMENTS

(Adapted from Carle and Olson 1985)

Indexed by Band (in order of decreasing electrophoretic mobility)

Band	Chromosome	Specific identifying hybridization probe
1	I	CDC19
2	VI	SUP11
3	III	SUP61
4	IX	SUP17
5A	VIII	ARG4
5B	V	URA3
6	XI	URA1
7	X	URA2
8	XIV	SUF10
9	II	LYS2
10A	XIII	SUP8
10B	XVI	GAL4
11A,B	XV,VII	SUP3, LEU1
12	IV	SUP2

DEFINING STRAINS

Strain	Resolved doublet	Genotype
AB972	(Standard)	*MATα trp1*
A364a	5A, 5B	*MATa ura1 lys2 ade2 ade1 his7 tyr1*
YPH45	10A, 10B	*MATa ura3-52 lys2 ade2 trp1-Δ1*

MAPPING STRAINS

Strain	Genotype
YPH80	*MATα ura3-52 lys2 ade⁻ his7 trp1-Δ1*
YPH81	*MATa ura3-52 lys2 ade⁻ trp1-Δ1*

YPH149 and YPH152 are derived from YPH80 and YPH81, respectively, via two chromosome-fragmentation events. They have the following differences from the standard karyotype:

Band 11 equals chromosome XV only (no VII).

A chromosome fragment (URA3⁺) derived from chromosome VII and carrying sequences centromere-proximal to RAD2 migrates between bands 10b and 11.

A chromosome fragment (TRP1⁺) derived from chromosome VII and carrying sequences centromere-distal to RAD2 migrates below band 1.

YPH149 and YPH152 grown in minimal medium (selecting URA⁺) show a more in-tense 90-kb chromosome fragment.

REFERENCE

Carle G.F. and Olson M.V. 1985. An electrophoretic karyotype for yeast. *Proc. Natl. Acad. Sci.* **82:** 3756–3760.

Strains

Experiment I	Looking at Yeast Cells
1-1 TSY623	*MATα ade2 his3 leu2 ura3*
1-2 TSY807	*MATa his3 leu2 lys2 ura3*
1-3 TSY800	*MATa/α ADE2/ade2 his3/his3 leu2/leu2 lys2/LYS2 ura3/ura3*
1-4 TSY481	*MATa/α ADE2/ade2 his3/his3 leu2/leu2 lys2/lys2 ura3/ura3* [pTS568]
1-5 TSY514	*MATa his3 leu2 lys2 ura3* [pTS592]

Experiment II	Isolation and Characterization of Auxotrophic, Temperature-sensitive, and UV-sensitive Mutants
2-1 S288C	*MATα mal gal2*
2-2 D665-1A	*MATa*

Experiment III	Meiotic Mapping
3-1 DDY700	*MATa, ura3-52, leu2-3,112, arg4-ΔBglII, trp2, cyh2*
3-2 DDY701	*MATα, ura3-52, trp1-289, arg4-Δ42, ade1,*
3-3 610.6D X 611.10D	
3-4 AAY1018	*MATa, his1 (same strain as 9-3)*
3-5 AAY1017	*MATα, his1 (same strain as 9-2)*
3-6 DDY702	*MATa, arg4-Δ42, his3Δ1, trp1-289, ade2*
3-7 DDY703	*MATα, arg4-Δ42, his3Δ1, trp1-289*
3-8 DDY704	*MATa, arg4-ΔBglII, his3Δ1, trp1-289*
3-9 DDY705	*MATα, arg4-ΔBglII, his3Δ1, trp1-289*
3-10 NE29	*MATa, trp1-289, ura3-52*
3-11 NE30	*MATα, trp1-289, ura3-52*
3-12 DRM117.39C	*MATa, trp2, leu2-3,112, ura3-52, arg4ΔHpa, rad3*
3-13 DRM117.126A	*MATα, trp2, leu2-3,112, ura3-52, arg4ΔHpa, ilv1-92*

Experiment IV	Mitotic Recombination and Random Spore Analysis
4-1 TSY812	*MATα can1 hom3 leu2 lys2 ura3*
4-2 TSY813	*MATa ade2 his1 lys2 trp1*

Experiment V	Transformation of Yeast

5-1	TSY623	*MATα ade2-101 his3-Δ200 leu2-3,112 ura3-52*
5-2	TSY502	*MATa his3-Δ200 leu2-3,112 lys2-801 ura3-52 tub4-34*
5-3	TSY1017	*MATa his3-Δ200 leu2-3,112 trp1-1 ura3-52*
5-4	TSY808	*MATa lys2-801*

Experiment VI	Synthetic Lethal Mutants

6-1	2405	*MATa ade5 his3 leu2 ura3 bub2::URA3 cyh2 [pMCM90]*
6-2	2406	*MATα lys2 his3 leu2 ura3 bub2::URA3 cyh2 [pMCM90]*

Experiment VII	Gene Replacement

7-1	BY4732	*MATα his3Δ200 met15Δ0 trp1Δ63 ura3Δ0*
7-2	2404	*MATα ade2 trp1 leu2 ura3 his3 lys2-801 ndc10::TRP1 [pRG68]*
7-3	2124	*MATa his3Δ1 ade2-1 ade6-1 hom3-H1 trp1-289 leu2-3,112 ura3-52 bar1::KAN cyh2 can1*

Experiment VIII	Isolation of *ras2* Suppressors

8-6	1784TRP	*MATα RAS2 his4 ura3 leu2 can1*
8-7	AMP141	*MATa ras2-530::LEU2 his4 ura3 leu2 trp1 can1*
8-8	AMP142	*MATα ras2-530::LEU2 his4 ura3 leu2 lys2 can1*

Experiment IX	Manipulating Cell Types

9-1	NE2	*MATα, ura3-52, leu2-3,112*
9-2	AAY1017	*MATα, his1*
9-3	AAY1018	*MATa, his1*
9-4	YSC006	*MATα ura3 ade2-1 trp1-1 can1-100 leu2-3,112 his3-11,15[psi⁺]GAL⁺*
9-5	YSC005	*MATa ura3 ade2-1 trp1-1 can1-100 leu2-3,112 his3-11,15[psi⁺]GAL⁺*

Experiment X	Isolating Mutants by Insertional Mutagenesis

10-1	2124	*MATa his3 ade2 ade6 cyh2 can1 hom3 trp1 leu2 ura3 bar1::KAN*
10-2	2194-12-2	*MATα his3 leu2 trp1 ura3 cdc28F19::TRP1 GAL-MPS1::URA3 can1*

Experiment XI	

11-1	Y190	*MATa gal4 gal80 his3-Δ200 trp1-901 ade2-101 leu2-3,112 lys2:GAL-HIS3:LYS2 ura3-52:GAL-lacZ:URA3* Cyhʳ

Counting Yeast Cells with a Standard Hemocytometer Chamber
(A. Kistler and S. Michaelis)

Using a **10x** microscope objective (and a 10x ocular), the "large" square shown in the circle below will fill your field of view. This square traps a volume of 0.1 µl. Therefore, to calculate the number of cells/ml in a particular culture, use the following formula:

cells/square x 10^4 x dilution factor = # cells/ml in your culture

STANDARD HEMOCYTOMETER CHAMBER

Modified, with permission, from Sigma-Aldrich Co. (copyright 1994)

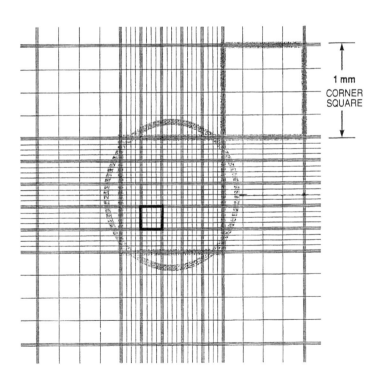

1 mm
CORNER
SQUARE

A general guide to typical cell counts for yeast cultures:

Culture	OD_{600}	Cell count	Dilution to count with hemocytometer
Sat'd YEPD	25	3×10^8/ml	10^{-2} or 10^{-3}
Log YPD	0.25	3×10^6/ml	10^0 or 10^{-1}
Sat'd SC-LYS	5	1×10^8/ml	10^{-1} or 10^{-2}

Generally, it is optimal to use a **40**x microscope objective, in which case only a portion of the circled field is visible; i.e., several "small" squares (a small square is indicated at lower left above). Then, count the cells in 5 small squares and use the following formula:

cells in 5 "small" squares x 5 x 104 x dilution factor = # cells/ml in your culture

Tetrad Scoring Sheet

(see next page)

Strain/Genotype									
A									
B									
C									
D									
A									
B									
C									
D									
A									
B									
C									
D									
A									
B									
C									
D									
A									
B									
C									
D									
A									
B									
C									
D									
A									
B									
C									
D									
A									
B									
C									
D									
A									
B									
C									
D									
A									
B									
C									
D									

Trademarks

The following trademarks and registered trademarks are accurate to the best of our knowledge at the time of printing. Please consult individual manufacturers and other resources for specific information.

ABI	The Perkin-Elmer Corp.
Bacto	Becton Dickinson & Co.
Coulter Counter	Coulter Corp.
Calcofluor	BASF Corporation
Eppendorf	Eppendorf-Netherler-Hinz GmbH
Kimwipe	Kimberly-Clark Corp.
Mg HotBead	Lumitekk
Sorvall	Kendro Laboratory Products
Stratalinker	Stratagene
Super Glue	Pacer Technology
Teflon	E.I. DuPont deNemours and Co.
Wizard	Promega Corp.
Whatman	Whatman International Ltd.
Zymolyase	Kirin Brewery (distributed by Seikagaku America Inc.)

Suppliers

With the exception of those suppliers listed in the text with their addresses, all suppliers mentioned in this manual can be found in the BioSupplyNet Source Book and on the Web site at: http://www.biosupplynet.com.

If a copy of BioSupplyNet Source Book was not included with this manual, a free copy can be ordered by using any of the following methods:

- Complete the Free Source Book Request Form found at the Web site at:
  http://www.biosupplynet.com
- E-mail a request to the info@biosupplynet.com
- Fax a request to 516-349-5598